T0138349

Stations in the Field

Stations in the Field

A History of Place-Based Animal Research, 1870–1930

RAF DE BONT

The University of Chicago Press

CHICAGO AND LONDON

RAF DE BONT is assistant professor of history at
Maastricht University in the Netherlands and lives in Leuven, Belgium.

The University of Chicago Press, Chicago 60637
The University of Chicago Press, Ltd., London
© 2015 by The University of Chicago
All rights reserved. Published 2015.
Printed in the United States of America

24 23 22 21 20 19 18 17 16 15   1 2 3 4 5

ISBN-13: 978-0-226-14187-9 (cloth)
ISBN-13: 978-0-226-14206-7 (paper)
ISBN-13: 978-0-226-14190-9 (e-book)
DOI: 10.7208/chicago/9780226141909.001.0001

Library of Congress Cataloging-in-Publication Data

Bont, Raf de, 1977–author.
Stations in the field : a history of place-based animal research,
1870–1930 / Raf de Bont.
pages ; cm
Includes bibliographical references and index.
ISBN 978-0-226-14187-9 (cloth : alk. paper)—
ISBN 978-0-226-14206-7 (pbk. : alk. paper)—
ISBN 978-0-226-14190-9 (e-book)
1. Biological stations—History—19th century. 2. Biological stations—
History—20th century. 3. Zoology—Fieldwork—History—19th century.
4. Zoology—Fieldwork—History—20th century. I. Title.
QH321.B66 2015
590.72—dc23
2014023201

♾ This paper meets the requirements of ANSI/NISO Z39.48-1992
(Permanence of Paper).

To my parents

Contents

INTRODUCTION

"Modern zoological research [. . .] aims to study the animal in its own dwelling place."[1]

The epigraph above comes from an article of 1905 by the German plankton specialist and former science journalist Otto Zacharias. With "dwelling place" [or: *Wohnplatz*] Zacharias was referring to the animal's natural habitat. It was by physically locating oneself in this habitat, Zacharias claimed, that the modern zoologist would find answers to his or her most pressing questions.

More than a hundred years after this statement was made, Zacharias's claim might sound counterintuitive. When we think of sites of animal research that symbolize modernity, the first places that come to mind are *not* the surroundings of the bird's nest, the octopus's garden in the sea, or—to include Zacharias's example—the parts of inland lakes in which freshwater plankton reside. The period around 1900, after all, witnessed the rise of grand urban research institutes that housed industrial-looking laboratories filled with mercury pumps, new-fangled microscopes, galvanometers, electric centrifuges, gas motors, contact clocks, and spectrometers. The plankton specialist who, despite the rise of these urban "engines of experiment," claimed that truly modern research was to be sought in distant woods, lakes, and seas, must, so it seems, have been an isolated eccentric.[2]

In fact, Zacharias was not an eccentric, nor did his ideas develop in isolation. He belonged to a much wider group of zoologists who were establishing a novel, indeed modern way of studying nature in the field. They propagated what present-day ecologists describe as "place-based research." Such research aims at general knowledge through detailed understanding of a specific place. It focuses on complex entities of interacting organisms, usually through studies over long periods in a natural field context.[3] Such a research does not approach the field as a place to merely collect and inventory species. Rather, it aims to study how organisms interact with each other and their environment by closely scrutinizing one place in particular. This approach was modern indeed, and, as such, it also needed a modern infrastructure.

Zacharias had straightforward ideas about the accommodation his type of science needed. As early as 1891 he had founded a biological field station near plankton-rich lakes in order to enable his ideas of place-based research. His strategy, thus, consisted in bringing the "dwelling place" of the scientist closer to that of the animals he was studying. And, again, this was not an exceptional idea. From the 1870s onward, a great number of biological field stations had been founded—first in Europe, later also in the rest of the world. They offered the context in which thousands of zoologists would receive training and perform research. Despite this importance, these stations and the practices they generated have received relatively little scholarly attention.

This book focuses on the early history of biological field stations and the role these played in the rise of zoological place-based research in the late nineteenth and the early twentieth century. It will explore the material and social context in which field stations arose, the actual research that was produced in these places, the epistemic claims that were developed there, and the rhetoric strategies that were deployed to convince others that these claims made sense. In short, it will study the intricate activities that enabled the zoologist to perform science in the animal's "dwelling place."

This book is itself an exercise in place-based research. It does not attempt to give a complete picture of the rise of the field station as a global phenomenon. Its focus is Europe (where the first stations originated), and even within this geographical scope most attention will go to a limited number of case studies. The highly localized character of the work performed in most field stations justifies an in-depth study of a small number of selected places. Just as the turn-of-the-century zoologists hoped to understand animals by observing them in their own milieu, we hope to understand turn-of-the-century zoologists by framing them in their natural habitat.

OUTHOUSED BIOLOGY

In many respects biological field stations constitute odd places of scientific research. They are built *in* nature (or at least close to it) while at the same time offering biologists a place to retreat *from* nature. Field stations provide researchers with their own scientific habitat, within the larger habitat of the organisms they study. Thus, they facilitate biologists both going outdoors and bringing organisms indoors. As indicated, this type of organizing biological research—although seemingly obvious—has relatively recent origins. Until the 1870s, zoologists and botanists had studied the living world in natural history museums, zoos, botanical gardens, urban laboratories, and of course in the field itself, but they had hardly created research institutions in or near the natural environment of their objects of study. Once the new idea caught on, however, the success was imminent. In his PhD thesis of 1940 the American biologist (and later social activist) Homer A. Jack counted no less than two hundred seventy of them, spread over the five continents. Not only for their numerical importance, but also for their apparent practical and theoretical innovations, the field stations earn their place in the historiography of science. Jack, for his part, was confident when he stated that the biological field stations "have loomed large in the progress of biological instruction and research."[4]

Compared with the botanical garden, the museum, and—most notably— the laboratory, the biological station has received a rather limited interest in academia.[5] Its rise to prominence in the life sciences came not much after that of the modern university laboratory and among historians of science it has stayed very much in the latter's shadow.[6] To be sure, quite some literature has been devoted to the histories of the most remarkable biological stations, but these are generally integrated in the traditional narrative of "the rise of the lab." In this narrative, it is a limited number of early, big and well-equipped marine stations that take the center stage: Anton Dohrn's Stazione Zoologica in Naples (Italy), the Marine Biological Laboratory at Woods Hole (United States), and the Plymouth Laboratory of the Marine Biological Association (United Kingdom).[7] It was particularly for their indoor work that these stations have become renowned. If field studies were performed in such stations at all, so it has been claimed, it was rather in the framework of student training than that of novel research. Purportedly, work in the field had a low status among late-nineteenth-century biologists and only smaller stations, entirely devoted to teaching, would have had a real focus on studies in the outdoors.[8] The overall image that arises from the existing scholarship is that, as far as research was concerned, biological stations were extensions of the urban laboratory—that is, places where organisms were attacked with microscopes, microtomes, mercury pumps, and kymographs, rather than explored in their natural habitat. Historian David

Allen was voicing a common image of the marine station when he stated that these "were really only outhoused laboratories, close to nature merely in the sense that they were sited on the very margin of the sea."[9]

If one delves deeper into Homer Jack's list, however, one easily finds research practices that do not match this general image. Many stations (whether devoted to the study of marine, freshwater, or land animals) would develop quite different research traditions than the prevailing ones in the laboratories of Naples or Woods Hole. Several of them gave rise to a fruitful tradition of fieldwork and they prided themselves on it. Some stations might have stuck to indoor work, but others invested a lot of time and energy in place-based research. Only by integrating this (often groundbreaking) work can we get a full picture of the history of the life sciences as they developed in the late nineteenth and the early twentieth century.

To accurately understand the appearance of the biological field station, we should strive for a broad and inclusive definition of what such a station is. I propose to use the term for every institution for instruction or research in the life sciences that located in (or next to) the field. In the decades around 1900 such institutions went under various names, defining themselves sometimes as "field laboratories," at other occasions as "zoological institutes," "marine stations," or "biological observatories." Often, such terms were used interchangeably to refer to the same place. Some of the late nineteenth and early twentieth century stations were broadly conceived, while others focused on one particular subdiscipline. Some oriented toward "pure" science, others to education or applied research. Some were set up by universities, others by local or national governments, scientific societies, or private initiators. Several focused on laboratory work; many others particularly drew field enthusiasts. Yet dividing them in strict categories would not work. Most, if not all, biological stations were true hybrids, mixed in their institutional origins, financial resources, scientific goals, research practices, and composition of visitors.

If their location in nature is the defining characteristic of biological stations, it is worthwhile to shortly indicate what "in nature" actually signified for the scientists who set up these places and worked there. The natural places in which stations were located ideally were regions in which a great variety of (undomesticated) organisms could be observed and/or collected alive. Such places were not necessarily devoid of humans, human activity, or human infrastructure. Rather, the opposite was the case. Station biologists often settled down in areas with long-standing traditions of fishing, hunting, and agriculture, or on sites with important tourist activity. The interaction of station biologists with these activities, and the ways in which the concept of "nature" was construed through this interaction, is a theme that will recur in this book.

NATURALISTS, OLD AND NEW

This book focuses on zoologists and animals, rather than botanists and plants. The history of turn-of-the-century field botany is rather well known, thanks to a sustained interest for plant ecology with historians of science.[10] Although some individual plant ecologists performed research at biological stations, the discipline as such seems to have largely developed in other working contexts. All in all biological stations were far more important for turn-of-the-century zoologists, who—in the words of the American limnologist Chauncey Juday—"deserve the chief, if not the entire credit for the founding of practically all of them."[11] These zoologists have attracted little attention so far. One of the probable reasons for this is that they have been written out of history by the later animal ecologists. In his pioneering *Animal Ecology* of 1927, the Briton Charles Elton stated that, up to then, field zoologists were mostly swamped by the collection "mania" of old-fashioned natural history. In the rare case that they did incorporate more experimental methods, they rather slavishly borrowed them from the plant ecologists, Elton argued. He had a straightforward explanation for this: "Botanists [. . .] finished their classification sooner than the zoologists, because there are fewer species of plants than of animals, and because plants do not rush away when you try to collect them."[12] Yet when we look more closely at what happened in field stations between the 1870s and the 1920s, we find much more than Elton's copycats and collecting maniacs. In various places innovative methods were developed to study animals in their natural environment.

The scientists who developed these new methods did not belong to one well-delineated discipline. Overall, one could argue, the boundaries between the different life sciences were highly unstable in the period around 1900. This is among others the case for the border between zoology and botany. The research project of the aforementioned Zacharias, for instance, might have focused on planktonic animals, but could hardly exclude the study of planktonic plants—or, for that matter, the physicochemical properties of the water in which both lived. Overall, place-based research in the life sciences was typically carried out by researchers from professionally heterogeneous backgrounds.[13] This book therefore focuses not on one discipline (zoology), but rather on a research object (the interaction of animals with their environment) that mobilized people from various backgrounds.

Within this heterogeneous group of scientists who worked at biological stations, several researchers made strong claims about the "experimental" character of their endeavor. These claims of experimentalism, and the ways in which these were translated into scientific practice, are an important topic of this book. In the period under discussion, the actual meaning of the term *experiment* was a highly contested subject. In this book it is therefore used as an actor's category (and a volatile one at that). Some laboratory biologists stressed—like many would today—that an experiment involved the manipulation of an independent variable by a scientist, and this in a highly controlled setting. These life scientists, of whom the French physiologist Claude Bernard served as a figurehead, believed that one well-controlled experiment could give access to universal laws. Simultaneously, they placed experiment in opposition to observation. The first was supposed to require active intervention and creativity, the second was presented as a passive, registering, and less prestigious activity.[14] Yet, such ideas and the epistemic ideals they contain were not shared by all life scientists of the period. This book will argue that station zoologists who defended place-based research held different epistemological standards. They also aimed for general knowledge based on induction, but they hoped to attain this by comparing observations of multiple, situated, complex phenomena in the field. This required larger networks and other techniques than those of the urban laboratory. Their research could, but not necessarily did, involve manipulation. The station zoologists might have claimed to be performing experimental science, but they surely gave this another meaning than it had in Bernard's physiology laboratory in Paris.

The discussions over the exact meaning of *experimentalism* were waged in a period that has received a lot of attention in recent scholarship. Counter to the traditional textbook story, several authors have lately claimed that the rise of the laboratory in the late nineteenth century did not lead to an eclipse of natural history, but rather to a revival and transformation of it. Eugene Cittadino has indicated how in the 1880s and 1890s German plant ecologists incorporated laboratory methods and ideals in their studies of plant adaptation, in this way drastically changing both the status and goal of botanical fieldwork. Geographically, these German university-based (proto-)ecologists largely focused on the tropics. They partially performed their studies "in the wild," but often they also made use of the botanical station of Buitenzorg in the Dutch East Indies—which, at that point, was the only one of its kind.[15]

Robert-Jan Wille's recent work has added the insight that the introduction of laboratories in the station of Buitenzorg offered academic botanists an inroad into applied agricultural science. He indicates that the role of the station in extending the professional prospects of these botanists was crucial to its success.[16] This might also explain why, in the early twentieth century, the model of Buitenzorg was readily copied by American scientists in the Greater Caribbean—as Megan Raby has described.[17] Yet, despite the global success of the Buitenzorg model, its botanist visitors were hardly involved in the rise of stations in Europe itself. The latter project, after all, preceded the former. Furthermore, it was more a regional than an imperial enterprise, and—as indicated—it was dominated by zoologists rather than botanists.

Whereas Cittadino, Wille, and Raby have explored the interaction between the botanical laboratory and the field in colonial settings, Robert Kohler addressed a similar theme with regard to the life sciences in the United States. In his inspiring (and somewhat controversial) *Landscapes and Labscapes*, Kohler conceptualized a "border region" between the cultural spheres of the laboratory and the field—a border region in which ecology and evolutionary biology would eventually flourish. This region, so he states, was opened up in the 1890s by a group of self-declared "new naturalists" who started to use experimental methodology for the study of organisms in their natural environment. Kohler indicates how biological stations as well as vivaria and biological farms were instrumental in the program of this "new natural history." Obviously, Kohler's innovative analysis of the lab-field border is in many ways very helpful to understand what happened in a similar period in Europe. At the same time, however, my book takes a different angle. Its focal point is not a presupposed border between two types of scientific culture, but rather a specific type of scientific workplace and the practices that developed there.

The focus on Europe brings a scientific world in sight which, according to various historians, differed drastically from that in the United States. Kohler has suggested that the "border culture" he described was a typical American phenomenon, hindered by more strict disciplinary boundaries in Europe.[18] Other differences often referred to by historians concern matters of patronage, social mobility, and the relative importance of theory and applied research.[19] Finally, the physical aspect of the field itself radically differed between the two continents. As a whole, Europe was much more urbanized and much more densely populated than the United States and this had strong historical roots. According to many historians, as a consequence of this, the European perception of the landscape was less shaped by the "wilderness" concept that was purportedly so central to the American mind.[20] Furthermore the differences in landscapes would have led to a less "powerful sense of disjunction between man's and nature's worlds" among Europeans.[21] It remains to be researched, however, whether these generalized differences hold true for the microcontext of the biological station and, if so, what they implied for its functioning.

Next to national differences, the social array of naturalist research complicates the picture. The lead figures in both Cittadino's story on German plant ecology and Kohler's narrative of American border science are mostly academic scientists, many of them laboratory-trained. Yet definitely not all turn-of-the-century researchers who were interested in the interplay between organisms and their environment shared this background. This becomes eminently clear in Lynn Nyhart's recent monograph *Modern Nature*. In this book Nyhart opens up the world of the German "practical natural historians": taxidermists, zookeepers, museum curators, and school teachers who largely operated outside the realms of academic science. Nyhart compellingly shows how this group was crucial in the development of what she calls "the biological perspective" (or: the approach "which viewed the organism as a living being embedded in nature, whose survival depended on its ability to interact successfully with both its physical environment and the other organisms around it"). In Germany, Nyhart claims, the roots of what later would be called "ecology" were populist rather than academic.[22] In substantiating this claim, Nyhart touches upon the role of biological stations only in passing. Obviously, these stations were often not "pure" examples of populist natural history, since many of them had ties with *both* academic biology and the realms of the practical naturalists. In my view, this makes them all the more interesting. Biological stations arguably constituted the most important meeting place between these two worlds in the period around 1900.

As indicated, the culture in which practical natural history flourished was contemporaneous with a culture that propagated zoology as an academic profession. The coexistence of these cultures led to cooperation, but also to tensions and a continuous renegotiation of scientific hierarchies. Historians of science (including Samuel Alberti, David Allen, and Jeremy Vetter) have, over the last decade, shown an increasing sensitivity for the history of "amateur" and "professional" participation in late-nineteenth and early-twentieth-century science. They agree that the rift between the two groups largely was a strategic rhetorical construct of this period, and that it resulted from a refashioning of roles in the scientific field in general.

They have further indicated that, in reality, the dichotomy hid a great variety of identities that were anything but closed, fixed, or homogeneous.[23] The biological field station, again, offers an understudied place in which these various identities came together, clashed, and were reshaped in the decades around 1900.

Using Peter Galison's terminology one might describe biological field stations as crucial "trading zones" in which researchers of various backgrounds could come together in order to work out a common language.[24] The negotiations concerned the type of knowledge that was to be produced, the kind of scientist that was needed to produce it, and the ways in which the latter had to approach the landscapes surrounding the station. In short, the negotiations involved the epistemic, social, and spatial aspects of nature study. These aspects will be at the center of attention of this book.

CASE STUDIES

To understand the role of a new scientific workplace for the actual science performed there, one needs to look closely. The "spatial turn" in the history of science has sharpened the historian's eye for the often very specific local conditions that play a role in the creation of knowledge.[25] These conditions varied drastically if we compare the biological stations of around 1900. The stations differed strongly in their social composition, leadership, geographical location, and self-representation. And although the researchers working there often strove for universal scientific insights, location also constituted a key element in substantiating their knowledge claims. Several stations indeed opted for place-based research rather than striving for the aura of "placelessness" associated with the laboratory. If we want to take this situatedness of knowledge seriously, micro-analyses will be more helpful than a broad generalizing story. For this reason, I have chosen to base this study on a limited selection of case studies. In this way, I will attempt to "go native" among the inhabitants of the early biological stations and to unravel their relations with their social and geographical milieu.

The case studies are selected in a way to give a fair representation of the variety of the stations at the time. To begin with, they include the disciplinary specializations that would get most connected with this type of scientific workplace: marine biology, limnology (or the study of inland waters), and ornithology. Furthermore, they comprise stations of varied institutional origins, showing differing engagements of universities, local governments, natural history societies, and museums. The leaders of the chosen stations present a similar variety. Among them, there is one influential university professor, one academically trained science entrepreneur, one former science journalist, one hunter-turned-pastor, and one museum curator.

The focus of this book will be on the German- and French-speaking science community in Europe—which both played a pioneering role in setting up biological stations. As such, the geographical focus will be on some of the most urbanized and industrialized areas of Continental Europe. At first sight, this might be an awkward choice. After all, fieldwork is often associated with unspoiled natural scenery rather than with the "artificial" landscapes of nineteenth-century Europe. Yet, although field stations were by definition erected "in nature," they were also unthinkable without a larger urban context. The institutional origins of biological stations—whether initiated by universities, museums, natural history societies, or local governments—were urban and most of their visitors were city dwellers. The relation with the city was, although often ambiguous, always of the greatest importance. Researchers explicitly retreated from the city, but not without staying connected. Next to material connections (in the form of railroads, for example) there were cultural ones. Most of the scientific ideals behind the field station were, after all, urban as well. Leaders of biological stations often distanced themselves from the purportedly one-sided indoor work in urban laboratories. But we will see that, at the same time, they also took these as a model. Even nature, as it was studied by the visitors of field stations, was largely an urban product. It was the urban bourgeoisie that would define *nature* as a place of scientific excursion and lonely contemplation, of beauty and health, of travel, tourism, regional pride, and economic value. Researchers in field stations incorporated some of these ideals, reacted against others, but certainly took them into account. Given this context, the choice for case studies in "artificial" landscapes can be advantageous. Their particular situation can help us to highlight the intricacies and ambiguities of working in nature in a larger world that is rapidly urbanizing.[26]

The scientists discussed in this book were obviously confronted with a particular type of nature. In a period when scientific explorers in the tropics still faced a natural world that could be considered wild and threatening, it was nature itself which was increasingly under threat in late-nineteenth-century Europe. With urbanization the pressure on natural resources increased. Because of overfishing, oyster and fish stocks declined, while populations of several bird species dropped—among other reasons because the feather industry needed plumes for the fashionable hats of metropolitan women. Rivers were dammed and increasingly polluted. Swamps were drained, fens colonized, and heathland brought under cultivation. The growth of international travel went together with invasions of foreign animal species (such as Colorado beetles or zebra mussels), which in many cases further harmed already damaged ecosystems. Agriculture and forestry more and more relied on intensive, commercial monocultures, which went together with ever more comprehensive policies of pest control. While large-scale capitalism gradually took over the landscape, remaining places of "picturesque beauty" became increasingly accessible for the growing number of tourists. For these, inland spas, beaches, and lakeside resorts would be engineered. Contemporaries usually perceived all these sweeping changes optimistically as being part of a conquest of nature, but in several milieus there was also a sense of loss. The latter feeling would among others fuel the foundation of the first associations for nature protection.[27] Naturalists who were stationed in nature itself were obviously close witnesses of the environmental changes mentioned. Often they would use this privileged position to explicitly profile themselves in the debates about how to actually administer nature in an urbanizing world—both as an economic resource and a haven for the soul. These debates were all the more intense in the countries that observed the most drastic effects of urbanization.

The case studies included in this book are selected not only for their location within the heartland of European urbanization, however. They were also chosen for their individual influence within the scientific world. Not every station that will be dealt with in this book was a traditional success story in the sense that it accumulated important discoveries. Yet all of them did well in propagating themselves as crucial places for the study of animals, and all of them managed to make specific locales into reference points in the world of science. They did so mostly independently from each other. This book, therefore, does not tell the story of a close-knit network. However, I do believe it is helpful to think of the apologists of biological field stations as belonging to one movement. The *station movement*—to coin a new expression—congregated researchers who promoted the study of nature from a permanent residence *in* nature. Although diverse in their particular research agendas, they all lobbied for an infrastructure that would enable them to study nature as much as possible in its situated complexity. The station movement thus can be seen as a counterpart to the *laboratory movement*—a term both contemporary scientists and present-day historians have used to describe the turn to the (urban) laboratory in late-nineteenth and early-twentieth-century science. This does not mean the two movements should be considered as opposites. This book will argue that with regard to methods, instruments, and ideals their relation involved both strategic distancing and partial appropriation.

So far, a historical overview of the rise of the late-nineteenth and early-twentieth-century station movement in Europe is missing. Chapter 1, therefore, sketches the movement's major developments and puts these in a wider context. It does so by exploring how turn-of-the-century biological stations related to the other scientific workplaces of the time. The hybrid character of the biological stations is addressed by discussing their relation with university laboratories, public aquariums, and natural history museums and by comparing their practices with the excursions of naturalist societies and large-scale state-sponsored surveys. Obviously the ways in which these different influences were integrated highly varied from station to station. The diversity indeed ranges from highly technological and indoor-oriented marine laboratories to poorly equipped wooden cabins in the woods.

One of the biggest, best-equipped, and most influential biological stations of the late nineteenth century was definitely the Stazione Zoologica in Naples, founded by the German zoologist Anton Dohrn. This iconic station, which combined a public aquarium with research facilities, has been rightfully celebrated for its ground-breaking laboratory work in evolutionary morphology and physiology. In his original program, however, Dohrn also included studies on the interaction of sea creatures with each other and their natural habitat. Chapter 2 explores this (often overlooked) aspect of his program and the ways in which Dohrn tried to realize it. The chapter, furthermore, describes how the spatial and social make-up of the station eventually hindered the translation of Dohrn's plans into concrete practice.

Chapter 3 zooms in on a station that could not have been more different from Dohrn's Stazione. It focuses on the small chalet at the beach of Wimereux (in France's Pas-de-Calais) that served as the marine laboratory for the French zoology professor Alfred Giard, his pupils, and his friends. Although starting out from a program similar to Dohrn's, the station in Wimereux generated very different practices. It became a major center for Giard's type of ecology, or, in his definition: "the science dealing with the habits of living beings and their relations, both with each other and with the cosmic environment."[28] This science, so he claimed, relied on the study of experiments that were prepared by nature itself. Chapter 3 unpacks this notion of nature's experiments—and explores how both the physical landscape in Wimereux and the moral landscape of Giard's station were involved in making this notion a successful one.

Although performed in Wimereux, Giard's science was obviously developed to be of value outside of its place of origin. And indeed his scientific approach would travel. Not only did it inspire fellow zoologists, but also it was applied in completely different working contexts. Because Giard's environmental determinism fit well in the ideology of the Third Republic, his work received political support and intellectual acclaim. Furthermore, echoes of his scientific approach (which centered on studying organisms in their environment) could be heard in paleontology and museology, agronomy and applied fishery research, sociology and psychology. As such, Giard's case offers a good starting point to explore how place-based research could transcend the site where it was performed and set a transdisciplinary enterprise in motion. Chapter 4, therefore, zooms out and explores how the highly localized projects of field stations could reverberate in wider scientific and cultural circles.

The examples of Wimereux and Naples illustrate the extent to which marine biology profited from the rise of the field station. The discipline gained academic respectability and was taken up by an increasing number of researchers. It is not so surprising then that life scientists active in other fields tried to copy its example. Limnologists were the first to be inspired by the successes of marine biology. Following Naples's model, the aforementioned Otto Zacharias founded Europe's first station for freshwater biology in 1891 in Plön (Schleswig-Holstein). In so doing, he hoped to generate enthusiasm among academics for a topic that so far had mostly interested only practical naturalists. The lake, so he claimed, offered a microcosm that would help researchers tackle wide-ranging biological problems. At the same time, his lakeside station was a microcosm in its own sort of way, putting on view the often painful interaction between academics and nonacademics in the turn-of-the-century life sciences. Chapter 5 will deal with this interaction and the ways in which it determined the early years of limnology.

Just like the limnologists, the ornithologists would use the foundation of field stations to raise the prestige and academic credibility of their science. Chapter 6 will illustrate this by focusing on the history of the ornithological observatory in Rossitten, East Prussia. This observatory, the first of its kind, was founded in 1901 and was led, for the first three decades of its existence, by the minister, hunter, and practical naturalist Johannes Thienemann. Initiated by the German Ornithological Society, it would quickly become an influential place that rejuvenated the study of bird

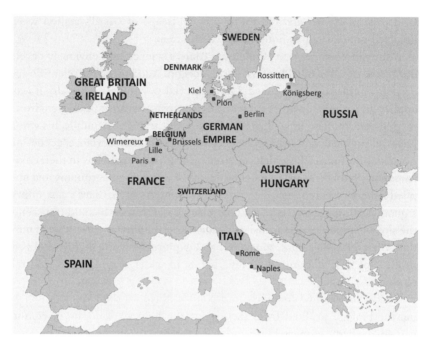

FIGURE 0.1 Location of the stations that serve as case studies in this book.

migration. The innovations that Thienemann introduced in ornithology called for specific spatial strategies. These included an optimal use of the natural characteristics of his workplace and the mobilization of a large network of geographically scattered amateurs. At the same time, his work altered the space he shared with the birds—materially, conceptually, and culturally. Thienemann's station proved a crucial tool in changing both ornithology and Rossitten.

Biological stations—whether for marine zoology, limnology, or ornithology—were often presented by their creators as modern enterprises. In this, they might seem unlike late-nineteenth and early-twentieth-century natural history museums, which, up until this day, carry a reputation of dusty, old-fashioned centers of accumulation and cataloguing. In fact, the histories of both types of workplaces crossed and field stations played an important role in the development of (some) museums in the early years of the twentieth century. Chapter 7 will substantiate this claim by focusing on the Natural History Museum of Brussels. In the first decades of the twentieth century this museum set up a cluster of biological stations, which were mostly ill-equipped and often of a temporary nature, but which proved decisive in the transformation of the institution from a place of classification and exhibition into a self-declared "exploration museum." Just like the stations in Wimereux, Plön, and Rossitten, the field stations of the Brussels Natural History Museum played an important role in putting the study of nature *in* nature on the agenda of the local scientific community.

Whether organized as annexes to museums, universities, civic societies, or public aquariums, biological field stations certainly offered a new workplace for the zoologist. This was not only of practical importance. The station movement enabled new questions to be asked, new approaches to be developed, new scientific identities to be constructed. It helped zoologists to broach new ground and broaden the scope of their science. Creating permanent workplaces in the animal's habitat, so this book argues, was instrumental in generating new scientific practices, developing new theories, and mobilizing new networks.

CHAPTER ONE

Stations and Other Places

When, by the late nineteenth century, the biological station had become a familiar workplace for life scientists, the urge grew to write its history and to celebrate its founding fathers. This enterprise quickly took on a patriotic flavor. The French were the first to claim fatherhood of the station movement. A gold-lettered inscription on the walls of the marine laboratory at Concarneau in Brittany, founded in 1859 by embryologist Victor Coste, claimed it was the first of its kind in the world. But the Belgians were quick to react. In a 1897 article, a Brussels science student bombastically proclaimed he had "the right and the duty to claim the glorious and fruitful initiative of marine zoological studies for Belgium." The student in question singled out Louvain zoology professor Pierre-Joseph Van Beneden as the *real* inventor of the biological station. He insisted that Van Beneden had set up a laboratory in the Belgian coastal town of Ostend as early as 1843.[1] Subsequently, historians have stuck to this foundation story.

Belgium's glory fades a bit when one takes a closer look at its earliest biological station. Van Beneden had at his disposal a small room in an oyster farm, containing "aquariums, fishing nets, tables, chairs, a bit of glasswork and a few reagents—that's all."[2] Before settling for Ostend, Van Beneden had tried other possibilities. With some effort he had managed to have living marine specimens sent to his cabinet at the university of Louvain, but quickly found out it was more convenient to take himself to the seaside than to have vulnerable marine invertebrates transported inland.[3] He investigated several locales, but Ostend was the only Belgian coastal town within easy reach of the newly opened railway. This was important. In the midcentury, railways played a similar role in opening up the coast for scientists and tourists in several other European countries. Yet, in the case of Van Beneden, it was not only the railroad that linked him to Ostend; in 1843 he married the daughter of a wealthy local tradesman. This was important as well. It was after all his well-to-do father-in-law, rather than his university, who financed his station. As with later biological stations, Van Beneden's marine laboratory was the product of private initiative and local alliances.[4]

Van Beneden's idea of a biological station found a following—somewhat hesitantly at first, but with growing enthusiasm after 1870. Next to marine laboratories, stations for the study of freshwater organisms and birds were eventually set up. This chapter will survey these developments as they occurred in Europe in the century after 1843, and frame them in the changing political, cultural, and scientific context of the period. In so doing, the chapter will explore the intricate relations of the biological station with other places of research. These places include permanent research infrastructures outside the field (such as the university lab, the public aquarium, and the natural history museum) as well as transportable devices used to survey the field (such as expedition ships and mobile laboratories). From these other places, the field station selectively borrowed technologies, practices, and intellectual endeavors, gradually creating something new in the process.

PIONEERS

In contrast to what later historians have claimed, Van Beneden's marine laboratory was not a complete novelty. After all, many late-eighteenth and early-nineteenth-century gentlemen-scientists had country houses, which they used as hubs to study locally available organisms. Some of these houses were located near the sea and some of them were rather well equipped. A good example is provided by the medieval-style castle of British gentleman John Stackhouse, elaborately described in a recent article by Anne Secord. The Stackhouse castle, built in the 1770s, had been set up on the Cornish coast so its owner could study seaweeds, which was a craze among the wealthy in late-eighteenth-century Britain. The castle's lower floors contained baths to study algal fructification and allowed Stackhouse to mimic marine tidal effects indoors in order to experimentally unravel seaweed procreation in seminatural conditions.[5] More than half a century later, Van Beneden would have envied such scientific luxury.

Despite some similarities, the stations of Stackhouse and Van Beneden should be framed in rather different contexts. The late-eighteenth-century

infatuation with sea organisms was fuelled by gentleman (and lady) collectors, primarily interested in decorative specimens such as algae. The
mid-nineteenth-century research, to the contrary, was largely driven by "professional" naturalists, who were equipped with compound microscopes and
dredges, and whose interest mainly concerned the biology of marine invertebrates and fish. The foundation of the station in Ostend was only one of
the many manifestations of this new interest. Van Beneden's project was,
after all, partially inspired by work across the borders. He knew of developments in England, where the British Association for the Advancement of
Science had set up a Dredging Committee in the late 1830s. He was familiar
with the work that Victor Audouin and Henri Milne-Edwards, professors at
the prestigious Muséum d'Histoire Naturelle in Paris, performed in a rented
farmhouse on Brittany's coast. And his friend, Berlin anatomy and physiology professor Johannes Müller, kept him well informed about the excursions that were launched from the North Sea island of Helgoland.[6]

Particularly Müller would prove influential for further events. He took a
keen interest in Van Beneden's station, which he visited in the revolutionary year 1848. As rector of the University of Berlin he had been confronted
with rebellious liberal-minded students, and hoped a retreat from the metropolitan epicenter of revolutionary activity to the coastal periphery would
calm him down.[7] And it did. In a letter to his son, Müller recounted how in
Ostend he could undertake embryological research "free of all emotions
and sorrows."[8] After a few months, he traveled to Marseille and Nice and
commenced what one of his biographers has described as his "years of sea
research."[9] He temporarily overcame his depression and converted a new
generation of students to "pelagic fishing." Several scientists who founded
marine stations in the following decades would explicitly refer to Müller's
agenda as the inspiration for their own enterprises. Often they shared his
ideal of research in retreat.[10]

The work of Van Beneden, Müller, and other "pelagic fishers" needs be
framed against the backdrop of the growing prestige that the study of marine invertebrates witnessed in the 1830s and 1840s. Although it is hard
to exactly pin down the reason for this rise in status, one might point to a
growing sense of the importance of marine invertebrates in the fossil record, and the fact that newly improved microscopes brought ever smaller
sea creatures within sight of the naturalist. Overall, the growing interest in
marine life did not seem to stem from a particular theoretical starting point,
nor did it cluster around one particular problem. Questions of distribution
and classification received a lot of attention, but so did physiological topics

such as salinity tolerance or blood circulation, themes concerning embryological development, or newly discovered and biologically astounding phenomena such as polymorphism (the existence of different morphological
forms within one species) and the alternation of generations (the existence
of different morphologies within one species, of which more than one can
reproduce). It was as a broadly conceived field that the study of marine invertebrates would capture the imagination of the leading representatives of
the profession.[11]

Men such as Müller and Van Beneden might have pioneered marine zoology, but overall they did not greatly invest in workplaces to perform this
study. Rather, they spent most of their time, energy, and money building up
their university museum collections.[12] Laboratories received less attention.
When working on the coast, Müller stayed in hotels and even his famous
Berlin laboratory was notoriously ramshackle. Van Beneden, then, might
have set up a fixed seaside workplace, but it had a provisional outlook from
the start. It was only rarely occupied and by no means equipped to house a
research institute. Although Van Beneden occasionally received guests there,
his station was primarily a private workplace. One of his early biographers
stressed that Van Beneden always spoke of "my dune laboratory" with "the
inflatedness of the proprietor."[13] His station belonged to the domestic sphere
rather than to the world of scientific institutions, and Van Beneden never referred to its existence in his publications. In the following decades, however,
the role, scale, and aura of the biological station would drastically change.

This shift was largely the accomplishment of one man: the aforementioned medical doctor Jean-Jacques-Marie-Cyprien-Victor Coste. Trained at
Montpellier, England, and Scotland, Coste quickly rose through the academic ranks in Paris and became the first holder of the chair of embryology
at the prestigious Collège de France in 1844. He set up an aquarium there
to study fish behavior, which grew famous among the Parisian beau-monde.
Political events enhanced his career. After the coup of 1851, he became personal physician to the new empress and close friends with Napoleon III.
Quickly, he would be involved in the emperor's bold plans for reforming
France. Mixing the rhetoric of science, utility, and conquest, Napoleon III
was eager to display his mastery over the French imperial landscape. He
appointed Baron Haussmann to reorganize Parisian urbanism, he furthered
plans to reforest large parts of the French countryside, and he strongly supported the foundation of a Jardin zoologique d'acclimatation, which aimed
to advance the management of animal resources through the acclimatization of exotic species. Finally, he hoped to domesticate the French waters

(and overcome food shortages) by setting up an "aquaculture" scheme, ap-plying agricultural techniques to ponds, lakes, and the sea. For this scheme he turned to Coste.[14]

France's big aquacultural plans originated in 1850, when a commission was created to study the possibilities of increasing fish "production" in French seas and lakes. Coste soon earned himself a place in the commission alongside established scientific heavyweights, such as Milne-Edwards and Achille Valenciennes. On the basis of a series of expeditions, he dramatically reported on the rapid decline in fish populations, and he called for massive projects of repopulation and artificial fertilization.[15] Coste had the ear of the government and in 1853 a station for aquaculture was opened in the Alsatian commune of Huningue. There, under the directorship of Coste, experiments were conducted to repopulate lakes, cross-breed different species, and ac-climatize exotic fish. Coste soon saw possibilities to extend his project to the seas. On one of his research trips he had encountered Etienne Guillou, a Breton fisherman, who, encouraged by the Naval Minister, had set up a state-sponsored fish-breeding tank in the port of Concarneau. Soon after meeting, Coste and Guillou decided to join forces. Guillou's tank was trans-formed in a "living laboratory" that officially opened in 1859. In the period of 1859–1861 alone it received no less than 400,000 francs in subsidies.[16] The resulting infrastructure was impressive. Six reservoirs were created by dynamiting the rocks in front of the laboratory building. Covering a total surface area of 1,500 square meters, they made up, as Coste put it, "a little miniature sea." Coste's laboratory contained sixty-eight aquariums on the ground floor, and a library and work stations for visiting scientists on the first floor. In Coste's own words the station constituted "a perfected research instrument, unprecedented in the history of science."[17]

The project raised enthusiasm, particularly among Parisian naturalists who admired Coste for his ability to mobilize resources for the life sciences. Armand de Quatrefages, professor at the Muséum d'Histoire Naturelle, was one of the many to praise Coste's perseverance. In a brochure, he stressed that thanks to Coste "we have seen three ministries enter with an unprec-edented generosity on the road of practical experimentation of the natural sciences, sciences that one day will be for the breeding of living beings, what physics and chemistry are for industry."[18] Well beyond his Parisian colleagues such as Quatrefages, Coste also reached members of provincial scientific societies and the general public with accessible publications and crowd-drawing aquariums at world exhibitions. But not everybody was en-thusiastic. In several circles, open disappointment was expressed regarding

A)

B)

FIGURE 1.1 A, a cartoon of Victor Coste "attending a thermal bath that suits him well" and B, an early depiction of his "vivarium-laboratory" in Concarneau. (Courtesy of the Wellcome Library, London)

the output of Coste's aquacultural projects and, more generally, the inflated egos of elite Parisian scientists.[19] Republican science popularizers condemned the expenses of Coste's projects, while maritime administrators characterized the embryologist as "a false prophet and a utopist." The notorious satirical newspaper *Le Charivari* ridiculed the aquacultural projects and caricatured Coste as an unworldly professor, taking baths in aquaria, teaching fish how to read, and feeding them until they grew so large that they started preying upon unsuspecting passers-by.[20]

Despite Coste's self-proclaimed successes, he remained very controversial. All in all, his prominence in French public debates and his connection to Napoleon III probably hampered the development of biological stations as much as they stimulated it. At least during the 1860s, no other stations were founded and Coste's "miniature sea" remained anomalous.[21]

THE CALL OF THE SEA

In 1870–1871 things changed. After the revolution in Paris and the war with Prussia, the Third French Republic came into being. In Versailles, the German Empire was proclaimed and Wilhelm I was crowned its first emperor. The newly founded Republic and the German Reich would both play leading, albeit rather different, roles in the take-off of the station movement in the following decades. The German Anton Dohrn set up the Stazione Zoologica in Naples, which turned into the most influential biological station of its day, but the French took the lead as far as quantity was concerned. In two decades the latter established no less than thirteen marine laboratories that geographically encompassed the country. The number would double over the next twenty years.

This sudden breakthrough does not seem to be instigated by radically new research interests. By and large the station zoologists of the 1870s built on the topics that had been en vogue in marine zoology for several decades. Of course, evolutionary theory gave an extra incentive on the promise that the stunning diversity of marine invertebrates held the key to new phylogenic insights. Yet the evolutionary viewpoint was not ubiquitous in the stations, and it integrated rather than overshadowed older traditions.[22] As we have seen, the idea of setting up stations preceded the success of evolutionary thinking. It seems in the end not so much new ideas that triggered the momentum of the station movement in the 1870s, but a series of institutional and cultural changes that were put to good use by a small number of well-positioned men.

In France, the initiator and the patron of the rapidly expanding station movement was the zoology professor Henri de Lacaze-Duthiers—a man of both great eloquence and determination. He was born a baron's son, and his career skyrocketed in the days of the Second Empire. He successively occupied chairs at the university of Lille, the Muséum d'Histoire Naturelle, and the very center of French Academia: the Sorbonne. Despite his early professional successes, Lacaze-Duthiers was not a Bonapartist and once the Third Republic was installed, he quickly showed himself a champion of the new spirit.[23] Hence, he explicitly framed his initiative to establish a marine station at Roscoff (in Brittany) as an attack on traditional "dogmatism" in French scientific instruction. Taking up the republican rhetoric, he stressed: "Science can only win by establishing true freedom, equal for all and therefore based on equality of working conditions appropriate to give rise to competition."[24] In concrete terms, Lacaze-Duthiers argued that each zoology chair should have access to its own well-equipped coastal laboratory. The Bonapartist aura that clung to marine zoology in Coste's heyday soon evaporated.

Lacaze-Duthiers gradually enlarged the modest station building he had established in Roscoff and turned it into a model of its kind in the French-speaking world. Although his teaching was centered in Paris, he explicitly presented Roscoff as "the country of my researches."[25] The concept of an extra-urban research station proved to be inspirational. Among others, Lacaze-Duthiers' students Alfred Giard and Edmond Perrier established their own marine laboratories in Wimereux (1873) and Saint-Vaast-la-Hougue (1888). Lacaze-Duthiers himself—annoyed by the fact that harsh Breton winters paralyzed scientific life in Roscoff—set up a second station at Banyuls-sur-Mer (1881), located on the milder French Mediterranean. By the turn of the century, most French coastal regions accommodated at least one marine station.[26]

In contrast to the French, German founders of marine stations were mostly active outside their homeland, particularly in Italy. German zoological activity there was the continuation of a long tradition. The *Italienfahrt* that had enthralled Goethe in the 1780s remained popular among the German intelligentsia throughout the nineteenth century.[27] As with Goethe, many nineteenth-century visitors combined an interest for Italy's art with a scientific eye for its nature. Johannes Müller, Carl Claus, Ernst Haeckel, Carl Gegenbaur, and Berthold Hatschek were only some of the naturalists who traveled in Goethe's footsteps.[28] The diverse and exotic marine organisms that could be fished from Italian shores stirred their aesthetic and scientific sensibilities and their descriptions inspired again other travelers. By mid-century, German zoologists knew which places to visit and which fishermen to work with.[29] In those years, the continuous influx of transient German zoologists had made a town like Messina into a "Mecca of the *Privatdozent*."[30] The biological stations set up in the Mediterranean only perpetuated and extended the traditional zoological *Italienfahrt*.

It was a small group of young entrepreneurial scientists who tried to set up marine zoology institutions on the far side of the Alps. In 1870 the director of the Berlin public aquarium, Otto Hermes, established a station in

Trieste, which mainly served for the collection and shipment of specimens to Berlin.[31] The first true marine station in the area, then, was founded by Hermes's acquaintance Anton Dohrn. An old student of Germany's leading morphologist Ernst Haeckel, Dohrn worked for a while as a *Privatdozent* (a poorly paid private lecturer) at the university in Jena. Eventually, Dohrn left academia to open the doors of his long-planned, large-scale, and internationally oriented Stazione Zoologica in Naples in 1872. The eight "working tables" he rented to foreign governments and scientific institutions were quickly filled with visiting researchers, and Dohrn gradually expanded the scientific infrastructure of his imposing neo-Renaissance station building. By 1910, there were no less than fifty "working tables" to be rented in Naples and the station—staffed and equipped beyond compare—had become an icon for the biological community worldwide.[32] Not all similar attempts proved as successful. Nikolaus Kleinenberg, for instance, another former Haeckel student, failed to replicate Dohrn's success in Sicily's Messina. Appointed to the chair of zoology at a local university in 1878, he tried to raise enthusiasm and money for marine research, but his project did not gain momentum. Kleinenberg's "station" remained for the most part a one-man-show.[33]

An aura of travel and retreat surrounded the field station and was part of its success. Since many were located in scenic surroundings, scientific research and professional networking could go hand in hand with a relaxed holiday atmosphere. In contrast, cramped urban zoological laboratories were notoriously badly accommodated at least until the 1870s. Field stations provided their more healthy counterpart. In 1883, an anonymous British author, envious of continental marine stations, wrote in *Nature*: "The success of these [marine] laboratories is doubtless increased by the fact that they are always in a healthy locality on a bracing seashore, so as to allow a realization of the apparently anomalous combination of work and rest. The scientist, worn out by fatiguing researches made in town laboratories, finds fresh elements of health and a fresh field for research by passing three or four months at a seaside laboratory."[34]

Both Dohrn's and Lacaze-Duthiers' initiatives inspired others to catch up with the movement that was spearheaded by France and Germany. In the 1880s and 1890s, the British and Scandinavian fishing industries supported the establishment of stations, which were understandably oriented toward applied research and often associated with local hatcheries.[35] The German fishing industry too became increasingly involved in the station movement and exerted pressure to supplement the Stazione in Naples with a marine laboratory located within the Reich.[36] In 1888 the powerful German Fishery Union set up a mobile laboratory for the zoological inventory of the Ems, Wese, and Elbe estuaries. Fishery-minded Wilhelm II took notice and formed a commission (which represented the Fishery Union, German government, and Prussian Academy of Sciences) for the establishment of a better-equipped permanent laboratory. The commission settled on Helgoland as the site for the new station. This small island in the North Sea had been a favorite destination of German excursionists for a long time, but only in 1890 would the German state acquire it from the British (by exchanging it for Zanzibar). The station opened in 1892 and served as a symbol of both territorial and scientific conquest.[37]

The institutional context of the development of German biological stations greatly differed from that of France. In general, late-nineteenth-century French scientific infrastructure noticeably lagged behind that of its neighbor and setting up stations was part of France's politically inspired catching-up project. Republican ideologues insisted that science was to be the guiding light to regenerate French society, and the scientists echoed their rhetoric. The French may have been defeated in the Franco-Prussian war, but many claimed they would overcome their fate by investing in science. With a sense of drama Lacaze-Duthiers stated in 1872: "The revival of the intellectual movement in France is in our eyes assured. It finds its origin in our defeat. It will be without limits, as have been our disasters and our calamities."[38] Crucial for this revival, he believed, were field stations. They became symbols of national resurgence, new ethics of sobriety and progressive science.[39]

In post-1870 France, zoologists and republican leaders were allies in their attempts to set up the so-called Republic of the professors, and, once the republicans came into power, the Ministry of Public Instruction immediately freed money for the organization of zoological stations.[40] The fact that in 1884 the influential post of director of higher education was taken by Louis Liard, a former student of Lacaze-Duthiers, obviously helped.[41] Departmental and municipal governments provided additional finances as did the newly established French Association for the Advancement of Science [*Association Française pour l'Avancement des Sciences*]. The latter was set up in 1872 after the British example to counter the centralization of science in Paris and to regenerate French society as such. Its motto was "through science, for the fatherland." Marine researchers came to dominate the zoology section of the Association and did not refrain from leveraging their position to obtain subsidies for their stations.[42]

French universities also contributed financially to the organization of marine laboratories, but they only did so to a limited degree. In many cases, stations were privately owned by their directors and not formally incorporated into the university system. Yet, since directors often held university chairs, their stations soon developed strong university ties, and became important sites of student training and PhD research. This was the case not only at the Sorbonne, which could make use of the two stations set up by Lacaze-Duthiers, but also at the provincial universities of Lille, Caen, Montpellier, Marseille, Lyon, and Bordeaux. These opened stations in Wimereux, Luc-sur-Mer, Sète, Endoume, Tamaris, and Arcachon, respectively. As local centers of science, these stations partially fulfilled the decentralizing mission of the French Association for the Advancement of Science, without financially burdening the universities. As main fundraisers, the station directors acted as dominant father figures. Some directors, such as Lacaze-Duthiers or Giard, invested their family capital in the enterprise, which only enhanced their strong personal involvement. Significantly, Lacaze-Duthiers was not buried in Paris, but was laid to rest in the garden of his station in Banyuls-sur-Mer.[43]

In Germany, the relation between universities and stations substantially differed compared with France. After 1870, German academia was considered to be an international model, but historians have also drawn attention to a growing internal crisis. The university population in Imperial Germany expanded, but did so unevenly. The number of increasingly powerful full professors lagged behind the general increase.[44] Junior staff members (extraordinary professors, *Privatdozenten*, and assistants) swelled in number, but were underpaid and excluded from decision-making university senates. A growing group of frustrated young researchers with poor career prospects was confronted with an exclusive "professorial oligarchy." This undermined scientific innovation and stifled institutional reforms. As a result, so the story goes, innovative science increasingly moved outside universities to hybrid institutes with mixed state and private financing.[45]

The story of German biological stations fits well with this narrative. Unlike in France, where marine laboratories were mostly established by high-ranking professors, it was less academically established naturalists who took the initiative in Germany. Dohrn was a *Privatdozent* with poor career prospects when he set up his station. Hermes was a university-trained chemist turned naturalist entrepreneur. And the first director of the Helgoland station, Friedrich Heincke, had a background as a primary school teacher.[46] In short, the stations set up by the likes of Dohrn, Hermes, and

Heincke lacked the personal links with university life that characterized French marine laboratories. Financially, they depended on governmental and industrial capital. Next to subsidies, which were voted on by national or regional Parliaments, gifts of scientifically minded industrialists crucially made ends meet. Anton Dohrn and other successful station directors habitually attended strategic social events so they could rub elbows with the rich and powerful. According to Dohrn, "dinners and suppers" were "the battle fields on which victories are won."[47]

ECHOES: LIMNOLOGY AND ORNITHOLOGY

Marine laboratories directly contributed to the scientific prestige and popular appeal of the study of marine fauna. Zoologists active in other domains quickly recognized this and eagerly emulated the formula. The first to do so were the limnologists. To be properly studied, freshwater organisms seemed to need a scientific infrastructure similar to their marine counterparts. Yet, students of freshwater fauna had less of a tradition to build on. Naturalists had long conceptualized lakes as "empty" spaces, largely devoid of life and zoologically uninteresting.[48] Only in the 1860s did the study of freshwater fauna gain momentum, when scientists discovered a surprisingly rich microscopic fauna in Scandinavian lakes. They were soon joined by Swiss physiology and anatomy professor Alphonse Forel, whose hydrographical and biological studies of Lake Geneva quickly earned him a European reputation.[49] It was the Germans, however, who were first to set up a permanent lakeside station. Zacharias's freshwater laboratory in Plön (1891) was modeled after existing marine stations and similarly funded by local and national governments, fishery societies, and private sponsors. Like many marine stations it also published its own journal: *Forschungsberichte aus der Biologische Station zu Plön*. Yet, as we will see in chapter 5, Zacharias struggled to make Plön the German center for freshwater plankton research and a national stronghold of limnology.

The laboratory in Plön was conceived as a place for basic research, but it was easier to raise funding for projects directly addressing fishery questions. Stations for applied freshwater research were set up in Friedrichshagen, near Berlin (1894), Silesia's Trachenberg (1895), and Munich (1898). Problems of urban and industrial pollution furthermore led to the establishment of the well-staffed Prussian Institute for Water, Soil and Air Hygiene in Berlin (1901). Zacharias's contemporaneous pleas to equally bolster the German infrastructure for "pure science" did not yield immediate results.

A. INTERIOR OF HOTHOUSE, SHOWING FLOOR BASINS AND MARGINAL AQUARIA.

B. INTERIOR OF COLDHOUSE, SHOWING MARGINAL AQUARIA, FLOOR, TANK, AND RACK.

CULTURE HOUSES AT LUNZ.

FIGURE 1.2 A hothouse and a coldhouse in the station for freshwater zoology in Lunz. From Kofoid, *The Biological Stations* (1910), 270.

Richard Woltereck, an extraordinary professor of zoology in Leipzig, found support for such plans only abroad. In 1906, he set up a limnological station in the Austrian town of Lunz, which, thanks to the patronage of a wealthy Viennese lawyer, soon gained a reputation as one of the best-equipped of its kind.[50] It impressively supplemented the natural variety of lakes and ponds in its vicinity with artificial pools, cement and glass aquaria, a library, darkroom, heating plant and greenhouses. The diversity in both natural and artificial "habitats" offered Woltereck and his students opportunities to experiment on heredity and variation. Woltereck further enhanced his station's reputation by founding the *Internationale Revue der gesamten Hydrobiologie und Hydrographie*, a journal that soon outranked Zacharias's as the leader in limnology. Nevertheless, in terms of intellectual aura and material facilities, the Lunz station was rather distinct and certainly not representative of European freshwater stations as such.[51]

The different institutional landscapes of France and Germany, so clearly reflected in the composition of marine laboratories, also marked the limnological stations. France's first freshwater station was set up in Besse-en-Chadesse (1893) by Paul Girod, then zoology professor at the university of Clermont-Ferrand. In its academic connection and self-proclaimed role in the decentralization of French science, it mirrored the French marine stations. Not coincidentally, it was patronized by the French Association for the Advancement of Science, and, following a French tradition, research at the station focused on acclimatization and repopulation of freshwater species. Local geography, with the nearby Puy-de-Dôme, invited research on the altitudinal distribution of plants and stimulated meteorological observation. Hence, a meteorological station and botanical garden were eventually added. Later, a fish-hatchery was attached. Following this example, other stations with fishery research agendas were set up by the universities of Grenoble and Toulouse, in 1901 and 1902, respectively.[52] Belgium followed soon after. In 1906 Ernest Rousseau (a zoologist with an obscure career path and poor health) established a freshwater station in a chalet in the small village of Overmeire. Despite links to the Brussels Natural History Museum, this was a private venture, and it struggled for survival before finally closing down in the 1920s.[53]

Overall, the modest limnological initiatives taken in Europe were fraught with common funding problems. As a result, the Americans quickly took the upper hand. In the 1890s several lakes and rivers in the American Midwest were endowed with their own stations, set up with support from state fishery commissions and inland universities that were situated too far from any oceanic coastline to establish marine laboratories. Pioneering ecologists (such as Stephen Forbes, Charles Kofoid, and Edward Birge) made them into an institutional and scientific success.[54] In 1905, Zacharias applauded the purportedly wealthy Americans for showing the way. "As a patriot and a German with a certain self-confidence," however, he also bemoaned that his compatriots were "dropping behind."[55] European limnology stations did, nonetheless, play an important role in the decades around 1900. They were crucial for both discipline-building and conceptual innovation in the study of freshwater fauna and, as we will see in chapter 5, for making lakes into important tools for solving general biological problems. According to freshwater zoologist Friedrich Lenz, "Limnological laboratories were there before limnology itself; they were the places in which this science came into existence."[56]

The fact that limnologists tried to emulate the material infrastructure of the more successful marine zoologists might not be so much of a surprise; they studied similar species in similar habitats with similar behaviors. More remarkably, ornithologists would be equally inspired. Although the practical side of research on birds seems to differ drastically from that of plankton or herrings, the first stationary bird observatories were explicitly set up as the ornithological counterparts of stations of aquatic biology. At the center of ornithological research, after all, was the observation of migratory birds, which echoed fish migration studies as performed in Helgoland.[57] Marine stations, however, were embraced by both the state and the fishing industry and thus raised money relatively easily. Ornithology, in contrast, largely remained the domain of civic naturalists who were forced to improvise with the limited financial means at their disposal.

Partially because of problems with funding, the first permanent ornithological station was set up only in 1900, but the idea dates back much earlier. At the founding meeting of the German Ornithological Society in 1875, pleas were already heard for research stations for the study of bird migration, and in 1884, at the First International Ornithological Congress in Vienna, a European network of temporary observation posts was developed. Subsequently, this generated an enormous amount of data on bird migration. Several national and international committees were set up to coordinate the enterprise. In the United Kingdom, lighthouses and lightships were commandeered to gather as much information as possible and in Austria-Hungary, foresters were called on to participate in migration research. It was hard, however, to actually link the various data generated by such projects and nobody was really engaged in processing the observations. One after the other the committees ceased to exist.[58]

Disappointing results prompted European ornithologists to change

FIGURE 1.3 Selection of the major marine, limnological and fishery, and ornithological stations in Europe around 1914 (indicated with squares, triangles, and circles, respectively).

tactics. Instead of a loose network of temporary posts, the German Ornithological Society decided, in 1901, to establish one permanently staffed station for observation, coordination, and accumulation of knowledge in a strategic place: the bird observatory of Rossitten in East Prussia. To ascertain the trustworthiness of the results, this new spatial strategy combined the use of a permanent station with a new research technique: bird banding. The Danish schoolteacher Christian Mortensen had introduced this technique in the 1890s, Johannes Thienemann adopted it in 1903, and thanks to him, the practice rapidly spread throughout Europe and America.[59]

The station of Rossitten, which will receive further discussion in chapter 6, served as an example. In 1909, a bird observatory was added to the biological station of Helgoland, which took up banding. Other ornithology stations were set up in Salzburg (Austria) in 1913 and Liboch (Bohemia) in 1914. The center of bird migration studies was clearly situated in Central Europe, and although British and French researchers launched banding projects, they never did so on the same scale as their colleagues in Germany and Austria-Hungary.[60] Even in Central-Europe, the number of permanent observatories remained limited, however, until at least the 1930s. As a rule they resulted from individual initiatives, often with the support of natural history societies and the state, but there was no academic discipline behind the enterprise. At that point, the ornithological community (apart from a single university-trained zoologist and occasional museum curator) consisted mainly of schoolteachers, foresters, military men, pastors, and wildlife artists, who were not professionally engaged in research.[61] Only later on did zoology departments develop an interest in ornithology and the work performed in observatories. There they encountered a world bustling with activity.

A HOME IN THE LAND OF EXPERIMENT

By the 1920s, European zoologists had built up a substantial infrastructure of biological stations at coasts, at lakes, and on major bird migration routes. These stations did not develop in a scientific vacuum. Their make-up was strongly influenced by the science performed at other scientific workplaces. These could function either as a model or as an antitype for what biological stations tried to achieve. Among these "other places" the urban-based university laboratory was probably the most important point of reference.

Historians frequently refer to the rise of the university laboratory and its epistemic culture of experimentalism as constitutive of late-nineteenth-century science.[62] In such accounts, the so-called laboratory revolution sets the stage for modern scientific research, including that of the life sciences. Robert Kohler, for instance, has argued that "the laboratory revolution made the cultural landscape in which biologists now live."[63] With such statements Kohler and other present-day historians echo nineteenth-century commentators, who were equally persuaded by the laboratory's revolutionary importance. American embryologist Charles Minot wrote in *Science* in 1884 that the laboratory was "the most remarkable and influential creation of science in our time." He added: "He who has never learned to appreciate a laboratory in its highest sense does not know even the meaning of 'I know.'"[64] For Minot, Europe, especially Germany, was the heartland of the laboratory revolution and he urged America to follow suit. In the United Kingdom, which was also regarded as lagging behind, Lord Kelvin reflected along similar lines. He wrote in *Nature* in 1885: "The search for the absolute and unmistakable truth is promoted by laboratory work in a manner beyond all conception." His conclusion: "No university in the world can now live unless it has a well-equipped laboratory."[65]

The question obviously arises as to how the zoological stations related to these highly praised university laboratories. Were they just a part of "the laboratory revolution"? Or was their foundation, to the contrary, a reaction *against* the scientific culture of the lab? Did they possibly constitute a middle ground between two cultures, and, if so, how?

Historians have recently suggested that "the laboratory revolution" —if ever there was such a thing—was an intricate and lengthy process, rather than the sudden appearance of a homogeneous type of workplace. In the words of Peter Galison: "there is [. . .] no single transtemporal, transcultural entity that is 'the laboratory.'"[66] The lab as a scientific workplace appeared many times and in many forms. In the nineteenth-century life sciences, the laboratory served various scientific disciplines and subdisciplines, which were themselves unstable entities. Yet it is clear that overall physiologists led the way. In the university of Breslau, Jan Purkýně had a laboratory space at his disposal from 1824, which is generally regarded as the first of its kind. Other physiology laboratories followed, in which small groups of elite students mostly worked under the guidance of just one professor. Yet only in the midcentury did the laboratory and its newfangled instruments really begin to play an important rhetorical and institutional role for physiologists. This was particularly true for a group of former students and acquaintances of Johannes Müller: notably, Carl Ludwig, Emil du Bois-Reymond, and Hermann Helmholtz. Known as the "1847 Group," these men increasingly distanced themselves from "descriptive" natural history by stressing the "experimental" character of their own science. In France, where physiology only gradually gained independence from medicine, similar claims were made by the likes of Claude Bernard. On both sides of the Rhine the laboratory as "a house of experiment" was presented as crucial for the identity of modern physiologists. It constituted, in Bernard's famous words, their "true sanctuary."[67]

For a long time, the experimentalist discourse of Bernard and the 1847 Group was only to a limited degree translated into brick and mortar. It took until about 1870 before well-equipped factory-like institutes for physiology were erected. Earlier research institutes that had been established from midcentury onward were comparatively modest, and not all countries were equally active in setting them up.[68] Prussia, for example, notoriously lagged behind. Around 1850 as influential a physiologist as du Bois-Reymond still experimented in his Berlin apartment.[69] In France, in the same period, the situation was worse still: "Most of our laboratories are in a miserable state," Louis Pasteur wrote in a confidential report to Napoleon III in 1868.[70] According to a nineteenth-century commentator even a celebrity such as Bernard had to perform his work "in a damp, small cellar—one of those wretched Parisian substitutes for a laboratory."[71] Despite these poor material conditions the experimental rhetoric of the physiologists both in France and the German lands conquered an important *intellectual* space. As such they could be threatening for practitioners in other disciplines. This was particularly the case for zoology.

When physiology started profiling itself as a nineteenth-century *Leitwissenschaft* [leading science], zoology was still in an early phase of discipline-building. Prior to 1850, only six German universities had an independent chair of zoology. In the others it was attached to anatomy or physiology in the medical faculty or incorporated into natural history in the philosophy faculty.[72] The situation was more or less similar in France. In 1855, the science faculties of only three French universities had professors for each of the kingdoms of natural history (minerals, plants, and animals). The others managed with two or, more often, just one.[73] The institutional alignment of zoology with natural history was matched with an intellectual one. In both France and the German lands, zoology often counted as the "descriptive" science par excellence: a discipline of classification and inventory performed in museums rather than laboratories. It was precisely this type of science that was increasingly belittled by the aforementioned group of physiologists.[74] Zoologists realized that if they wanted to emancipate their discipline they would have to transform its focus and aura.

In the late 1840s this transformation process was launched in the German lands with the foundation of a journal carrying the programmatic title *Zeitschrift für wissenschaftliche Zoologie* [Journal for scientific zoology]. The German example was followed two decades later by a group of French zoologists around Lacaze-Duthiers, who propagated a so-called experimental zoology. Medically trained morphologists led the way in both cases. They defended a broad program of zoology with morphological and developmental studies of marine invertebrates at its core and the microscope as its iconic instrument. In the German lands, the "new" zoology movement clearly preceded the rise of the biological station (and prepared the terrain for its success). In France, the two phenomena were chronologically closely intertwined.

The history of the clash between Bernardian physiology and the new zoology in France has been recounted several times.[75] Because this clash between two institutionalizing disciplines was of crucial importance for the rise of French biological stations, it is appropriate to sketch its outlines once more. The conflict—a "war" in the words of Lacaze-Duthiers—was set off by a report that Claude Bernard wrote in 1867 for the Minister of Instruction. It dealt with "the progress and the march of general physiology in France." In his report Bernard distinguished "contemplative" "observational sciences" from "explicative" "experimental sciences." The latter were thought uniquely capable of "conquering nature." Physiology was obviously presented as such a progressive science, whereas zoology, along with other forms of natural history, was considered merely descriptive. In his personal notes, Bernard stressed that "science can never have as its goal to heap labeled stones, dried plants, stuffed animals, pieces in jars." His actual report was more cautious in its phrasing, but equally stressed that extra state subsidies were particularly needed for the laboratories of first-class science, and far less so for the "observational" sciences.[76] Such conclusions could not but stir a lot of controversy.[77]

Victor Coste was the first to take up the defense of zoology. On the one hand he claimed to be grateful to experimental laboratory science, even presenting himself as one of its leaders, but on the other hand he argued that the "observational sciences" —including zoology—were as "conquering" and "explicative" as Bernard's physiology. He claimed that manipulative experiments were necessary only when "the secrets of nature were hidden from the eye": "Everywhere where our gaze can reach them, they do not need any artifice for forcing their organization to manifest itself." Rather than stressing the importance of laboratory manipulation, Coste emphasized the significance of localizing oneself *in* nature. Laboratories, he insisted, should not simply be tethered to urban university chairs: "They should be transported [. . .] to all places of this world where there are great problems to be solved, problems that can only be solved *there*." Concarneau obviously served as an example of one of these places.[78]

Lacaze-Duthiers entered the fray in 1872, the same year he founded his biological station. He attacked Bernard in the manifesto-style opening article of his own journal, *Archives de zoologie expérimentale* [Archives of experimental zoology]. His position was unambiguous: "Desirous to take part in the claiming of the legitimate rights of zoology, I would like to prove that this science can and should be experimental." At the same time, Lacaze-Duthiers contested Bernard's definition of what experimentalism actually was. Echoing the chemist Michel Chevreul, Lacaze-Duthiers argued that human manipulation was *not* constitutive of an experiment (or an "experience" in Chevreul's terminology).[79] In Chevreul's and Lacaze-Duthiers' view, the experimental method consisted of two steps. First a hypothesis about causal relations in nature was established (on the basis of observation or reasoning). Second, this hypothesis was tested by carefully and repeatedly observing natural phenomena. To Lacaze-Duthiers' mind, it did not matter methodologically whether these observed phenomena were manipulated by man or not. This implied that much of his own zoological work counted as experimental—as would most future research at other biological stations. At stake for zoologists at the time was, in the words of Lacaze-Duthiers, nothing less than "a home in the land of experiment."[80] The biological stations provided the material home, Chevreul's theory the conceptual one. For decades to come, the directors of biological stations in France would present their establishments as a counterproof of Bernard's conceptual mistake.[81]

Whereas French biological stations appeared in an antagonistic intellectual climate, the German scientific community was more relaxed on the matter. German founders of biological stations were also self-declared representatives of "scientific zoology," but they belonged to its second or third generation. Like their French colleagues, they defended a broad research program, which, next to morphology, focused on issues of classification, histology, generation, biogeography, and the effects of the environment on form and function.[82] Overall, they felt little need to distinguish their program from the more narrowly defined work of the 1847 Group. On the contrary, when looking for support for their undertaking they were eager to solicit it from this increasingly powerful generation of physiologists. When Dohrn, for instance, attempted to persuade the German government to hire working tables in his station he made sure to contact major members of the 1847 Group. Among others, du Bois-Reymond and Helmholtz reacted supportively.[83] Strategically for Dohrn, both were key members of the politically powerful Academy of Sciences in Berlin, a position they also used to back other station directors.[84] Thus, the institutional coming of age of German "new zoology" was in fact facilitated by the physiologists.

Of course, the biological station was not the sole architectural symbol of the new zoology's scientific status. Following the physiologists, zoologists also moved into prestigious city-based institutes, which housed teaching and research laboratories. In fin de siècle Germany, zoological laboratory buildings were erected at among others the universities of Leipzig (1880), Freiburg (1886), Berlin (1888), Würzburg (1889), and Heidelberg (1894).[85] Simultaneously, the Third French Republic freed money for similar construction works. The new nine-million-franc Science Faculty of the Sorbonne,

which opened in 1889, housed among others several laboratories for zoology. Other universities, such as Lille and Montpellier, erected analogous
(albeit less costly) constructions in the 1890s.[86] In Belgium's university of
Liège, then, Pierre-Joseph van Beneden's son Edouard opened his institute
for zoology in 1889. The newspapers scornfully described it as "the palace
of the beasts."[87] The transition, in Liège like elsewhere, was decisive. Earlier
urban university laboratories for zoology had been makeshift at best, while
most money available for the discipline was spent on building up natural
history collections. Now zoological laboratories were accommodated in spacious buildings, self-consciously designed as urban landmarks and adorned
with classical pillars and gables.

Yet the modern academic zoological institute did not compete institutionally with the biological station. A division of labor between the two
quickly took shape. A traffic of (living and freshly killed) specimens from
stations to university laboratories crystallized, while—particularly between
semesters—students and academic staff traveled in the other direction.[88]
This was not only the case in France, where station directors were by profession integrated in the academic system, but also in the rather different
institutional contexts of Germany, Belgium, and other European countries.
Both urban laboratories and field stations profited from the expansion of
European science faculties toward the end of the nineteenth century. In
these years, rather than being outshone by the academic institutes, biological stations confirmed their important role for the identity and the popular image of the profession. News on field stations was prominent in
elite periodicals such as Zoologischer Anzeiger, and even more conspicuous
in popular journals like La Nature. From 1875 to 1900, the latter journal issued fifteen elaborately illustrated articles devoted to zoological stations, but
completely ignored the new urban institutes of the zoology departments.

Biological stations always incorporated laboratories of some sort. The
term "laboratory" was often used interchangeably to refer to the station as
such. Yet biological stations boasted a particular kind of laboratory culture
that developed dialectically with that of urban-based university laboratories.
From the beginning, the leaders of the biological stations strove for a certain
kind of experimentalism, but without leaving aside the old naturalist interests. They tried to keep up with the specialization in their profession, but
without losing the intellectual breadth of the new zoology. And they tried to
provide "universal" truths, while stressing that many of the great problems
of zoology could not be solved in placeless urban sites, but only at specific
locations in "real" nature. These mixed ambitions characterized biological

stations in their early development and—as the following chapters will show—for several decades to come.

PLACES OF DISPLAY

The founders of the original biological stations were clearly inspired by the laboratory archetype. But the station movement also integrated ambitions, technologies, and practices from a variety of other places of science. Some of these places were sites of scientific display, popularization, and spectacle as much as they were venues of proper research. Particularly influential were the public aquarium and the natural history museum.

A British creation of the midcentury, the public aquarium appeared on the Continent alongside the first marine stations. The histories of both establishments are closely intertwined and the early success of the marine laboratory leaned on the concurrent aquarium craze that swept Europe. The craze originally concerned home aquaria and took off among the Victorian bourgeoisie. The trend was triggered by the combination of a sharp drop in the glass tax and a series of technological innovations. In the 1830s airtight glass-cases had already been devised in which microcosms of plants could flourish despite urban pollution. The following decade, chemists experimentally proved that the self-sustaining principle also applied to aquaria (with plants producing oxygen and animals consuming it). From 1850 onward, aeration and filtration apparatuses were introduced to optimize the equilibrium, handbooks started to appear, and aquarium shops opened their doors. The first of these, the Aquarium Warehouse of William Alford Lloyd near London's Regent's Park, would employ no less than fourteen collectors of marine animals to meet the demand.[89]

Whereas the home aquarium was within reach only of the prosperous, the public aquariums would parade the underwater world for all social classes. The first of these opened its doors at the London Zoological Gardens in 1853. The initiative was a popular success and in the following two decades public aquaria would be established in most of Europe's main cities. Lloyd became the foremost engineer of the movement, subsequently devising the public aquariums of the Paris Jardin zoologique d'acclimatation, the Hamburg Zoo, and the Crystal Palace in London. He integrated newly invented mechanical contrivances for adding atmosphere, and experimented with indirect light, putting his audience in the dark and having the aquariums lit up.[90] Despite integrating truly modern technology, these aquariums were designed to confront the public with a wild and hitherto hidden nature. As one early visitor of European public aquariums matter-of-factly commented: "In all cases an attempt has been made to keep from the mind of the visitor the idea that water pipes, pumping and blouse-wearing attendants—none of which withal are oppressively tidy—are necessary for the well-being of the tanks."[91]

One of the people to be impressed by Lloyd's engineering skills was Anton Dohrn, who visited the Hamburg aquarium in 1866. Dohrn was already hatching vague plans for setting up a zoological station, but the practical side of it kept worrying him. The idea of merging a station with an aquarium only arose three years later when, on a visit to Berlin, he again immersed himself in the world of underwater spectacle. He went to see the newly opened aquarium at Unter den Linden that combined Lloyd's technology with further artifice, among others assimilating the aquariums to manmade grottoes.[92] Since the Berlin establishment was highly popular Dohrn became convinced that the aquarium craze could further his project. He reasoned that the entrance fees of a public aquarium would actually suffice to sponsor the functioning of a research station. The idea materialized and Dohrn created a building of which the first floor housed the public aquarium and the second and third the research laboratories. Lloyd would supervise the construction of the aquarium, for which the technology was imported from England.[93]

Although the returns of the Naples aquarium remained below expectation, Dohrn's hybrid institution was copied elsewhere. In Europe, the stations of among others Sète (1881), Banyuls-sur-Mer (1881), Millport (1884), Plymouth (1888), Den Helder (1890), Endoume (1891), Port Erin (1892), and Helgoland (1901) would each eventually house its own public aquarium. In some places the research infrastructure would even be an addition to a preexisting aquarium—for example, in Arcachon (founded in 1882) and Le Havre (in 1883). Unlike metropolitan aquariums, the aquariums of marine stations mostly focused on regional fauna, combining a sense for the spectacular with a whiff of science popularization and localism. Banyuls-sur-Mer offers a good example. In its "exhibition room" aquariums with fauna from the French Riviera were alternated with busts of Lacaze-Duthiers' personal pantheon of naturalists; seats were present for popular lectures that could be illustrated by the station's stereopticon; and in the evening the aquariums were spectacularly lit by an arc lamp. One of the staff members stressed that the exhibition room had an ornate design with reason. After all, it served "as the reception salon, where lay persons could come to initiate themselves in the curiosities of zoological science."[94]

As a rule the public part of the station was architecturally closed off from the laboratory rooms, separating science from amusement through walls, stairs, and separate entrances. In publications of the time it was stressed that such arrangements made sure that researchers would not be disturbed by the clamor of visiting tourists. The disconnection was obviously also symbolically important in acquiring an aura of scientific professionalism devoid of hobbyist connotations. Yet the separation between the world of aquarist amusement and zoological research was only a partial one. In fact the aquarist technologies had found their way to the actual laboratories of the stations. So-called research aquariums belonged to the basic equipment of marine and limnological stations as much as microscopes and reagents. Their pumps, filters, and airing apparatuses had first served the excitement of the bourgeois household, but they worked equally well in the less prosaic world of the new zoologist.[95]

The interaction between marine zoology and entrepreneurial spectacle was particularly clear in the undertakings of the aforementioned Otto Hermes. In 1871 this chemist was appointed codirector of the Berlin Aquarium at Unter den Linden to become its sole leader two years later. The aquarium had a station to import Mediterranean species in Trieste, which not only served as a trading port but also as a research institution. Hermes made his geographically branched undertaking flourish, not in the least because he used his Berlin premises to stage great apes and anthropological shows. At the same time, he earned himself a name in scientific circles. His trade in living specimens, which were transported in "glass balloons," was highly valued among German academics, his research on the habits of the conger eel met with some acclaim, and his chemical procedure to artificially create seawater was positively referenced in scientific journals. The water, more importantly, was a significant source of income since it could be sold to other German (inland) aquariums. Because, unlike most directors of marine stations, Hermes was engaged in artificially recreating habitats in places far away from their original geography, he heavily depended on elaborate technologies such as artificial water, railways, and glass balloons. On a smaller scale, however, directors of biological stations relied on similar practices for bringing water creatures closer to the eye of the zoologist and the general public.[96]

It was exactly the same aim that would eventually involve the biological stations in the development of underwater photography. As Edward Eigen has shown, this new technology only gradually emancipated itself from the aquarium. In the 1890s Paul Fabre-Domergue—adjunct-director of the station in Concarneau—experimented with photographing so-called double

FIGURE 1.4 Modern technology hidden by atmospheric grottoes in the Berlin Aquarium in 1869. From Klös et al., *Die Arche Noah* (1994), 349.

aquariums that were set up to render "realistic" images of creatures in their "natural" habitat. The procedure included a foreground aquarium containing the pictured animals and a background aquarium adding an element of depth. Another strategy with similar aims was developed at the station of Banyuls-sur-Mer by Lacaze-Duthiers' former assistant Louis Boutan. Boutan was confronted with the problem that the group of "ear-shells" he studied was hard to keep in the station's aquariums. He therefore set up diving expeditions to study the species in situ. Eager to leave a "more or less exact description" of his underwater impressions, he then caught the idea of experimenting with photographic technology on the sea floor. He designed a series of underwater cameras and magnesium lamps, which would be constructed in the atelier of the station. His photographs, although initially set up for scientific purposes, were reproduced in popular monographs and displayed for a wide audience at the Universal exhibition of 1900. Their success both echoed and extended that of the public aquarium.[97]

Public aquariums and universal exhibitions were not the only places of display to impinge on the station movement. Another, albeit a less obvious influence, was the natural history museum. A few stations were actually set up by museum curators, often in an effort to free their institution from an association with purportedly old-fashioned projects of collecting. If the museum was to be more than a "necropolis," some argued, it had to have its "annexes" in living nature.[98] So, the Paris Muséum d'Histoire Naturelle— much maligned by the rising university zoologists—founded a marine station on the island of Tatihou in 1888.[99] In the early 1900s, then, the Brussels Natural History Museum associated itself with a marine laboratory, a limnological station, and a public aquarium.[100] Such examples were rather rare, however. More often than serving as an outpost for metropolitan museums, stations actually performed museological functions in the periphery. Marine, limnological, and ornithological stations were frequently provided with exhibition rooms that presented a survey of the local fauna. Despite the stations' aura of modernity, many of these exhibitions were in fact fairly traditional. By and large they existed of a few shelves with taxonomically arranged specimens, sometimes haphazardly extended with local antiquities. Most of the time only limited financial resources were freed for exhibition purposes.[101]

Two notable exceptions were the modernly conceived Oceanographical Museum in Monaco and the North Sea Museum of the Helgoland station. The museum-annex-station in Monaco, set up by the wealthy prince and amateur oceanographer Albert I, was the most lavish of the two. In *Nature* a commentator noted in 1910 that its inauguration festivities did "not fail to impress the most regardless pleasure-seeker in the gayest haunt of the Côte d'Azur with the thought that science, even in its least known department, was a thing of high importance."[102] Despite significant laboratory infrastructure, the museological function outshone research in Monaco. Visitors were impressed by the great entrance hall lit with electrical medusa-shaped lusters (inspired by the drawings of Ernst Haeckel) and the spacious exhibition rooms that housed broadly conceived displays. Among others these rooms contained ship models and scientific apparatuses, illustrating both the techniques of the marine industries and scientific oceanography. Next to taxonomic collections, they showed "biological assemblages" that explained the relationship between species and their environment.[103] The latter aspect was also prominent in Helgoland's museum, founded in 1898. Beside taxonomic collections, the North Sea Museum put "biological" problems on view concerning the life history, development, food habits, and behavior of the local marine animals, and, like in Monaco, assemblages were put up.[104] This type of display was not entirely new, but echoed modern modes of representation of metropolitan museums in both Europe and the United States.[105] More than most of these, however, the museums of Monaco and Helgoland used the assemblages to stress the importance of the scientific study of organisms in their local setting, thus highlighting the raison d'être of the biological station.

SITES OF SURVEY

The laboratory, public aquarium, and museum were in origin all urban creations. The biological station was crucial in transporting technologies and practices of these institutions to the field and adapting them to this new locale. At the same time, however, biological stations integrated central aspects of preexisting fieldwork, echoing the culture of surveys, excursions, and expeditions. In the integration process they would again drastically transform them.

The heyday of the biological station coincided with that of another scientific institution: the naturalist society. Although several learned societies dated back to the ancien régime, they got a second wind in the latter half of the nineteenth century. In France provincial societies profited from the same decentralizing tendencies that benefited the biological stations. Partially because of this reason, France remained unrivaled in its number of *sociétés savants*. Yet also in the German lands they became more prominent from the 1840s onward, and Belgium witnessed the same phenomenon a few decades later. Several of these societies established close ties with the newly set up biological stations. Some stations (e.g., in Arcachon or Rossitten) were actually cofounded by societies, while others (like Plön or Plymouth) received important funding from them. In still other cases the bond was not institutional or financial, but personal. The stations' staff, after all, often participated in the activities of the local societies, and individual amateur society members occasionally worked at the stations.[106]

Natural history societies and biological stations shared one prototypical activity: the excursion. By the time biological field stations arose, "excursioning" was an established naturalist tradition with its own equipment and prototypical landscapes to visit. Not coincidentally, biological stations were often founded in established destinations for both amateur and professional excursionists. Furthermore, the station's equipment would to an important degree consist of the dredges, nets, field notebooks, guns, portable microscopes, and dissecting gear used in traditional field excursions. Yet there was a shift in approach.[107] Whereas traditional excursions largely consisted of rapid surveys of the *unknown*, biological stations would mostly focus on a permanent monitoring of the same well-known place. The rationale of the latter strategy was not to discover ever more species, but to study the field and its organisms from a "biological perspective." The transition from the first type of activity to the second was a gradual one and, arguably, was never fully achieved. Initially, several directors of biological stations still had one foot in the old paradigm. Lacaze-Duthiers, for example, first intended to set up a "flying laboratory" that was to make an inventorying "Tour de France."[108] Otto Zacharias for his part had to convince himself (and his colleagues) that the territory in which he settled would not get scientifically exhausted in the long run. Zoologists with a main interest in faunistics continued to prefer so-called wandering stations, which could be moved once a place was thoroughly surveyed.[109]

The "wandering station" was not a mere abstraction for that matter. In Holland a wooden laboratory that could be taken apart and reassembled effectively traveled the shoreline from 1875 onward. It offered zoologists transportable aquariums and folding chairs. The Dutch sandy beaches, so the argument went, were too poor in species to justify prolonged stays of naturalists.[110] Not much later, the university of Aberdeen equipped a similar mobile laboratory to survey the fauna of the Scottish coast.[111] Wandering stations like those in the Netherlands and Scotland seemed an intermediary step in the spatial organization of zoology. This step was repeated every time new subdisciplines came under the spell of the station movement. In the late 1880s a flying laboratory for limnology surveyed the Bohemian lakes under the direction of Antonín Frič; a decade later, an ambulant station for ornithology and entomology was set up in Southwestern France; and around 1900, plans were made to design a "swimming laboratory" to study the fauna of Germany's main rivers.[112] Most of these initiatives were short-lived. After a while, the flying laboratories ceased to exist or were replaced by permanent stations. By adopting a place-bound biological perspective, zoologists more and more ceased their itinerant existence.

The fact that stations received a permanent character obviously did not imply that visiting scientists were stuck to one place. Mostly field stations were used as a hub for excursions in its surroundings. When distances were short, the landscape penetrable, and equipment not too heavy, these excursions were done by foot. In other cases trains, horse carts, and boats (ranging from canoes to well-equipped steam vessels) were employed. From the 1930s onward, cars were in use for expeditions over land. The limnological station of Lunz, for example, would furnish an Opel Blitz as a driving laboratory.

A)

B)

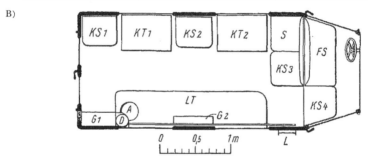

Abb. 3. *LT* Laboratoriumstisch. *KT* Klapptische. *FS* Führersitz. *S* Fester Sitzplatz. *KS* Klappsitze. *G* Verschließbare Gestelle für Reagentien und Glasgeräte. *D* Destilliertes Wasser. *A* Ablauftrichter. *L* Eiserne Leiter.

FIGURE 1.5 A, The "flying" laboratory of Antonín Frič and B, the "driving" laboratory of Franz Ruttner. The words in part B read "LT laboratory table, KT folding table, FS driver's seat, S fixed seat, KS folding seat, G closable shelf for reagents and glass instruments, D distilled water, A drain funnel, L iron ladder." From Frič and Vávra, *Die Thierwelt* (1894), 9, and Ruttner, "Ein Mobiles Laboratorium"(1933), 151, respectively.

Franz Ruttner, then director of the station, stressed that the time when the limnologist's equipment fitted in a single rucksack was over. Particularly for chemical analyses of the water the laboratory-car was deemed useful. The necessary equipment was, after all, too heavy to be carried to the lakes in question, while the chemical composition of the water was considered too temperature-dependent to be transported to the station. The car meant one could do the research on the spot.[113]

Cars, carts, canoes, and steam boats were used to study the coasts, lakes, and lands in the immediate vicinity of biological stations in the late nineteenth and early twentieth century. Contemporaneously, the study of the deep sea underwent its own kind of professionalization. From the 1860s onward, deep sea research had acquired increasing visibility thanks to heavily mediatized expeditions—and these expeditions would constitute yet another point of reference for the founders of biological stations. The expedition ships used for deep sea research can be seen as elaborate (and expensive) ambulant laboratories. They served as sleeping quarters, mode of transport, and scientific instrument.[114] The most iconic example of these ships was without doubt the British screw corvette HMS *Challenger*, which sailed the world between 1874 and 1876. It was the first ship to be sent out with the express objective of studying the ocean. A converted navy vessel, the *Challenger* was equipped with purpose-built laboratories, a dredging platform, and the most modern thermometers, photometric apparatuses, and sounding machines. The results of the Challenger expedition were generally esteemed and other nations joined the enterprise. The Germans, for instance, launched the high-profile Plankton expedition (1889) under the lead of Kiel physiologist Viktor Hensen and the German Deep-Sea Expedition (1898-1899) led by the Leipzig zoologist Carl Chun. Mostly, these expeditions combined an interest in zoology with a study of the physical and chemical properties of the ocean's water and floor. Often the zoological part focused on dredging up unknown species, preferably from great depth. Eventually, however, the ocean researchers took an increasing interest in the interaction between animals and their environment, which found its clearest expression in their rising interest for the problems of plankton distribution in the oceans.[115]

The marine station and the expedition ship rooted in a similar intellectual ground. Both profited from a popular enthusiasm for the seas, failing fisheries, and a growing nationalist spirit of scientific competition. Both types of scientific workplace were partially propagated by the same men—scientists such as Karl Möbius in Germany, Prince Albert I in Monaco, or Thomas Huxley in Great Britain. Yet only in a few institutions did the concrete practices of the station and the deep-sea expedition overlap. One of these was, again, the Oceanographical Museum of Monaco. This institution largely grew from the oceanographic cruises that Prince Albert I performed beginning in the 1880s. Although his expedition ship (and pleasure yacht) *Princesse Alice II* did not officially belong to the museum, the ties were very close. The first director of the station, Jules Richard, was also the laboratory leader of the *Princesse Alice II* and the personal secretary of the Prince.[116] A somewhat similar situation arose in Helgoland when, in 1902, the station became the seat of the German section of the International Commission for the investigation of the North Sea, and the station's staff started to use the well-equipped steamer of the Commission.[117] These were exceptions, however. Expedition ships were expensive and the study of the deep sea usually did not belong to the special focus of marine stations. Anton Dohrn, for example, would provide the expedition of the Italian corvette *Vettor Pisani* (1883–1885) with scientific advice, but he never set up such large-scale oceanographic expeditions himself. [118]

To be sure, there did exist a mutual transfer of technology and methodology between the zoologists based in biological stations and these venturing on deep-sea expeditions. Station limnologists, for example, would gratefully adapt the deep-sea dredges and samplers for their own use, just like they would appropriate the quantitative methodology developed by scientists like Hensen for the study of freshwater plankton.[119] At the same time, however, the station and the expedition ship developed quite differing practices and spatial strategies. Expedition ships were suited to collect, but far less to process, information. The ships' laboratories were usually dark and cramped, and, mostly, they were used to perform only the most urgent conservation procedures.[120] The actual processing of the data happened ashore, far from the place where the material was actually gathered. Identifying the species collected on expeditions, dissecting them, or counting them in samples often engaged several scientists and their inland laboratories for years, if not decades.[121] All this hardly counts as place-based research. Revisiting the same place, a common practice in biological stations, was virtually impossible for researchers interested in the deep seas. Of course station-based marine zoologists, limnologists, and occasionally even ornithologists also performed boat expeditions, but their territory remained limited. The station's sailboats, motor launches, and small steamers were used for studying well-defined regions of its the immediate surroundings. Unlike the oceanographic expedition ships, they hardly ventured into "unknown territories." Rather they stayed within their restricted local habitat.

Biological stations thus incorporated a hybrid set of technologies from laboratories, aquariums, museums, and ships. And one could even expand the list. Most stations, for instance, included their own library—thus assimilating the tradition of one of the oldest places of learning. It is true that station libraries were sometimes limited to a few shelves, but in other cases they housed ambitious reference collections. In 1910, the library of the Stazione in Naples consisted of 13,000 bound volumes, which were made accessible by card catalogues and specialized personnel. As did many other stations, the Stazione expanded the collection by exchanging its own journal with others, and by asking offprints of visiting scientists.[122]

In addition to libraries, photographic darkrooms were common in zoological stations. Photography (like aquarium keeping) had an air of novelty and encompassed the worlds of both middle-class entertainment and science. Thanks to its promise of objectivity the camera was, from the 1890s onward, taken up as a field instrument, employed by zoologists for inventorying the environment of the station and the changes it underwent.[123] It was employed, for instance, to inventory variation among fish, the location of bird's nests, or the borders of vegetational groups.[124] Furthermore, it was used to bring the invisible into the light. In this context, we already discussed the use of underwater photography in Banyuls-sur-Mer. We could add the use of microphotography for the study of plankton, which became popular in several stations from the 1880s onward; or the use of aerial photography, practiced since the 1920s, which was used to get a bird's eye view of the interaction between geology, hydrology, vegetation, and animals.[125]

Next to photo studios and libraries, still other places could prove inspirational. Some stations (such as the ornithological observatory in Rossitten) would keep and display live animals, in this way incorporating breeding practices of the zoo. Others echoed the techniques of the botanical garden by keeping herbaria (such as Naples, Helgoland, and Roscoff) or by constructing greenhouses (such as Lunz). Some stations (e.g., Banyuls-sur-Mer) included their own mechanic's workshop in order to develop and repair scientific instruments adapted to local needs. By integrating such workshops, biological stations could strive for a kind of autarky that, in the words of Lacaze-Duthiers, was useful if one was situated at "the end of the world."[126] Finally, station directors often installed classrooms and living quarters within the confines of their stations. In this way these would both architecturally and socially echo the atmosphere of nonscientific places such as schools and hotels. In short, the biological station was many places in one.

The mixed technologies of biological stations were modern in the sense that they incorporated new inventions developed for laboratories, expedition ships, or photographic studios. At the same time, most stations did not constitute a very high-tech environment. Airplanes and laboratory cars arrived relatively late and really were the exception. Also, the endless laboratory rooms and intricate equipment of the Stazione Zoologica or the limnological station of Lunz were certainly not characteristic for the movement as a whole. More typical were modest field microscopes, dredges, and binoculars. In many cases, both the technology and the architecture of biological stations carried an improvised character. Often the stations themselves were reused buildings (hatcheries, schools, chalets, or even painter studios), while instruments were simplified versions of tools transferred from other contexts. As we will see, station zoologists frequently took pride in their improvisation skills and the relative discomfort of their infrastructure. With some exceptions, it is clear that the modernity of the station movement does not lie in its use of sophisticated equipment. Rather it has to be sought, as we will see, in its novel epistemic claims, its mobilization of new networks, its improvised spatial strategies, and its interaction with quickly changing landscapes.

It might be clear by now that, despite their geographical retreat from the urban centers, biological stations did not develop in isolation. Rather, they echoed a diverse range of cultural practices and projects that circulated in late-nineteenth-century Europe. The biological station bore, among others, the marks of Napoleon III's ambitions to manage the nature of the Empire, of the Third French Republic's belief in a decentralized progressive science, and of the entrepreneurial spirit in the German Reich of the late nineteenth century. It was influenced by the leisurely tradition of the *Italienfahrt*, the heated mid-nineteenth-century discussions on "experimental science," the academic professionalization of the life sciences, and the revival of the natural history society. Stations were used to observe nature, to inventory it, to experiment with it, and to display it through artificial recreations. At the same time they served as the architectural symbol of the zoologist's expertise. Their mixed technology, borrowed from very divergent contexts, was used to tackle a wide range of questions. If the biological station carried a scientific program at all, it was broad and inclusive. And, as we will see, it came in many local variants.

CHAPTER TWO

Naples

Indoor Sea Creatures

It is impossible to write about the history of zoological stations without deal-ing with its most iconic representative: the Stazione Zoologica in Naples. As early as 1910 the Stazione was described as "one of the most potent factors in the development of modern biology."[1] By that time, indeed, both the rising stars and established mandarins of the life sciences had found their way to Naples. The station's visitors list reads as a *Who's Who* in turn-of-the-century biology. It includes names such as Edwin Ray Lankester, August Weismann, Robert Koch, Charles Otis Whitman, Thomas Hunt Morgan, Theodor Boveri, Paul Ehrlich, Hans Driesch, Jakob von Uexküll, Oscar Hertwig, and Jacques Loeb. But not only in the quality, also in the quantity of its visitors Naples easily outcompeted other stations. In the first thirty years of its existence it welcomed no less than 1200 life scientists, and, by the early twentieth century, it had become a cliché to refer to Naples as the biologist's Mecca.[2] The metaphor is telling. It indicates the central role the station played in the identity building of biologists, and it highlights the extent to which the place had become an obligatory destination for the ambitious.

Like contemporary scientists, present-day historians have presented the Stazione as the zoological station par excellence. In their analyses they have particularly focused on the part the station has played in the development and the spread of new forms of laboratory biology. Among both scientists and historians the Stazione is best known for studies in descriptive mor-phology, and later also for experimental work in embryology, cytology, and physiology. Its role as a site of international exchange of laboratory tech-niques has long been recognized. The Stazione may have been situated close to the animals' habitat, but this mainly served the purpose of providing its laboratory with fresh specimens. The Mecca of the turn-of-the-century biologist, one has to conclude, was a highly equipped lab, not a field site.[3]

If one wants to make the argument—as this book does—that the station movement played a crucial role in transforming biological work in the field, it seems that the Stazione is external to that enterprise. And yet, this chapter will argue that the aspirations of its founder, Anton Dohrn, were, to the contrary, typical to the station movement as a whole. What Dohrn hoped to develop (as did many other protagonists who figure in this book) was a broadly conceived zoology that concerned not only the animal's morphology, embryology, and physiology, but also its behavior and interaction with its environment. The latter was deemed as important as the former. Yet, despite Dohrn's efforts, the study of animals in their natural environment largely failed to materialize in Naples. Because of this reason, we have essentially forgotten about the breadth of his ambitions. In the historiography of science Dohrn's initial aspirations have become overshadowed by his eventual realizations.

The story of the increasingly influential laboratory research at the Naples station has often been told. This success story is historically relevant, but it obscures as much as it enlightens. It leaves out Dohrn's struggle to develop lines of research outside a laboratory context. This chapter focuses on that struggle. It highlights Dohrn's ambition to use his station for studies of animal life habits, and it looks into his efforts to materialize these studies. Finally, it offers explanations for why his efforts were largely in vain. These, I believe, have to be sought in the social and spatial logic of his station. It was material and organizational circumstances that eventually hampered the part of his program that went beyond the laboratory.

THE DOUBLE LEGACY OF DARWINISM

In 1872 Dohrn published two programmatic articles, in which he outlined the scientific agenda of the station that he was setting up in Naples. The articles, published in *Nature* and *Preussische Jahrbücher*, were part of a strategy to prompt scientific interest and stimulate fundraising. Their publication constituted the apotheosis of a campaign Dohrn had been carrying for two years. In this period he had been defending the idea of a zoological station in correspondence with well-known zoologists and politicians, in scientific gatherings and city councils, and in articles in the daily press in Britain, the German lands, as well as Italy. In *Nature* and *Preussische Jahrbücher*, then, he finally outlined the role his future station would play in the development of zoology as a whole. It is clear that Dohrn did not lack ambition. The Stazione, he claimed, would offer a new type of infrastructure for a new type of science.[4]

Dohrn's vision on both the future of science and its infrastructure had been gradually shaped by his experiences as a young aspiring zoologist in the previous decade. Dohrn, the son of a wealthy industrialist, received training in zoology at several German universities. It was his stay at Jena under Ernst Haeckel and Carl Gegenbaur that would prove to be the most influential. In 1862 Haeckel brought Dohrn in contact with the work of Charles Darwin, which stirred the young student to "piercing excitement."[5] In the following years it would also be a particularly Haeckelian version of Darwinism that Dohrn was to embrace. His interest—like that of Haeckel and Gegenbaur— was in the reconstruction of evolutionary trees on the basis of anatomical and embryological research. In 1863, Dohrn left Jena to finish his education in Berlin and Breslau, only to return two years later. On a coastal excursion with Haeckel in Helgoland in 1865 he decided to redirect his research from entomology to marine zoology. It was then that the first idea arose of setting up a permanent station next to the sea. Although both Haeckel and Gegenbaur had doubts about Dohrn's scientific capabilities, he was allowed to prepare his habilitation in Jena. In 1867 he became Haeckel's first (unpaid) assistant.[6] The reverberations of Haeckel's thinking would be clear in all his subsequent projects.

Dohrn soon participated in the Jena phylogenetic research program, developing his own morphological theories about the evolution of arthropods. In the context of this research, he organized repeated excursions at the Scottish coast, together with the naturalist David Robertson. "Bad weather" and "distemper" chased him from Scotland, however.[7] With a self-designed portable aquarium, he moved to Messina in Sicily, where he set up a small research laboratory. In the meantime, diverging views on scientific methodology, materialist philosophy, and vertebrate origins gradually estranged him from what he described as "the Haeckel-Gegenbaur firm."[8] Financial worries brought him temporarily back to Germany, where—as we have discussed in chapter 1—he developed his plans for a station-annex-aquarium in Naples. While setting up a place of his own and gaining international recognition, Dohrn lost much his former insecurity and he started an open attack on Gegenbaur's and Haeckel's theories of the origin of the vertebrates. The animosity between Jena and Naples quickly sharpened and would last a lifetime.[9] Be that as it may, Dohrn's articles in *Nature* and *Preussische*

FIGURE 2.1 Dohrn in a microscoping pose. (Copyright Stazione Zoologica Anton Dohrn di Napoli–Archivo Storico)

Jahrbücher show that the ideas of the Jena circle still echoed in his conception of zoology.

Darwinian thinking, Dohrn indicated in his programmatic articles, had thoroughly changed the questions relevant for the zoologist. To his mind traditional systematics (which centered on the classification of species) and faunistics (which focused on the listing of species within certain geographical areas) fell "very far short of the problems now ripe for solution." Darwin had brought bigger questions to the fore—questions that needed new instruments to solve them. According to Dohrn, post-Darwinian biology looked "pretty much like a boy who has suddenly grown in one year out of all of his clothes, presenting the ridiculous aspect of a man in a child's dress." The most urgent garment for his discipline's wardrobe, Dohrn argued, was a well-equipped zoological station.[10]

The first spearhead of Darwinian biology, according to Dohrn, had to be embryology. He claimed it was this discipline that now "carried the chief burden of progressing science." In a very Haeckelian fashion, but without mentioning Haeckel's name, Dohrn argued that Darwinian embryology was not just about accumulating facts, but about progressing ideas and principles. His main theoretical instrument for this was Haeckel's recapitulation theory, which stated that the embryonic development of the individual recapitulated the evolutionary history of the species. From this angle, embryological observations on primitive sea animals were crucial in tracing the "history of the animal world from its earliest beginnings." Dohrn's station then was to provide the infrastructure for overcoming the "mechanical difficulties of observation" that had thus far largely hampered embryological research on marine species.[11] He claimed that running water, aquariums, and laboratory equipment close to the animal's natural habitat would open up a whole new research agenda. This claim proved to be prophetic. Historians have elaborately discussed the successful embryological work that, since 1872, was deployed in Naples by both Dohrn and the many visitors to his station.[12]

Yet, embryology was only one of the "two great departments of biological science" that, according to Dohrn, went "ahead of all other." The second, maybe more surprisingly, concerned the animal's "habits and conditions of life" (or, in German, the *Lebensweise der Tiere*). Again, it was Darwinism that in Dohrn's view had given this field a completely new relevance. Haeckel and Gegenbaur might have neglected this aspect of evolutionary research, but it certainly had been a prominent interest of Darwin himself. It was the animal's habits and conditions, after all, that were central to the struggle for existence and natural selection. Through the lens of these concepts, old questions received new meaning. Firstly, Dohrn indicated, these questions touched upon the individual animal, its living space, food, and procreation and gestation habits. Second, it also involved the interaction between animals, which Dohrn described in terms of "war and peace," or the question of "eating and being eaten." Finally it concerned the conditions (*Bedingungen* in German) or the milieu (*Medium*) in which the animal was living: the pressure, temperature, and moisture level of the air; the geological composition of the soil; the influence of vegetation; and, for aquatic animals, the chemical composition, temperature, and currents of the water. All these elements, Dohrn indicated, could influence the habits of individuals, their morphology, and their survival rate. Thus, Dohrn's project on the "habits and conditions of life" concerned what present-day biologists would describe as "ecology." Dohrn used a then common expression among German life scientists: *Biologie im engeren Sinne* (or "biology in the narrow sense").[13] Despite its name, this project was far from narrow. At least in Dohrn's conception it included the behavior and "external physiology" of individual animals as well as the dynamics of animal populations.

Unlike embryology, which thanks to quickly evolving slicing and staining techniques was on the rise in academia, *Biologie im engeren Sinne* was
certainly no priority at the European universities of around 1870. According
to Dohrn, the topic was "rather out of fashion." In Britain, it was mostly
outside of the universities one had to look for students of animal life habits.
Dohrn mentioned Darwin and naturalist-explorers such as Alfred Russell
Wallace and Walter Bates. In German universities, then, he believed the
topic was on the wane as well. It was only a few isolated professors who
took up the subject: the physiologist Karl Theodor von Siebold in Munich,
the parasitologist Rudolf Leuckart in Leipzig, and the proto-ecologist Karl
Möbius in Kiel. Overall, however, Dohrn believed the *Lebensweise der Tiere*
was far from being a priority at the German zoology departments. He cyn-
ically pointed out that it had been a poet who discovered alternation of
generations, while a clergyman was the first to describe parthenogenesis.[14]
And, according to Dohrn, even among amateurs the interest for animal
life habits declined. In this way, what he saw as "one of the most urgent
necessities of biological study in our time" was left almost entirely without practitioners. Again, Dohrn represented his station as a possible solution to the problem, since it would bring the zoologist closer to the living
animal.[15]

In many ways, Dohrn's interest in animal life habits was rooted in an old
tradition of natural history. In a letter to a friend he had indicated as early
as 1866 that it was actually natural history and not zoology that constituted
his true "love."[16] In 1872, however, he was hesitant to use the term *natural
history* with its old-fashioned undertones. Rather than stressing the importance of tradition, Dohrn emphasized the modernity of his post-Darwinian
enterprise. It might have been natural history he was after, but he made sure
to indicate it was of a new and an ambitious kind. It was the same strategy
that later would be taken in an appeal to the Reichstag in which the leading
German scientists Hermann Helmholtz, Rudolf Virchow, and Emil du Bois-
Reymond requested that Dohrn's station be subsidized. In this appeal they
referred to a "natural history in the higher sense of the word."[17]

In short, Dohrn set out to create a Darwinian zoology with a twofold
object. The first was the laboratory-oriented project of reconstructing evolutionary trees on the basis of embryological research. The second was to understand the role of animal life habits in the struggle for life, which implied
going beyond the laboratory. Once the station was in use, however, the first
project quickly overshadowed the latter. One of the reasons for this has to
be sought in the location of the Stazione.

BIG CITY LIFE

Dohrn wanted to bring the zoologist closer to living nature. He decided, however, to set up his station in one of the biggest cities of nineteenth-century Europe. This choice obviously came with practical implications when it concerned the study of free nature. Dohrn, nonetheless, had good reasons for settling in Naples.

One crucial element in Dohrn's choosing of a site was his idea to make his station into a place of both scientific activity and tourist entertainment. The concept of combining a research lab with a money-generating public aquarium obviously made him dependent on sightseers for an important part of his revenue. In this respect Naples, with its strong age-old tourist industry, was a sound option. Ever since the eighteenth century European "men of taste" had come to the city, attracted by the adjacent volcano, the antiquities, and the Naples opera scene. Also the local color gradually became a source of attraction for northern holiday-makers. With the rise of Romanticism, the chaotic city and its population were increasingly described in orientalist tones as being fascinatingly vulgar and passionate.[18] Although the high days of Romanticism were over by the time Dohrn set up his station, he could still rely on a well-developed tourist infrastructure. All he had to do was make sure that his aquarium would be favorably mentioned in the most popular travel guides of the period, such as Baedeker's and Murray's.[19]

The metropolitan aspect of the Stazione's location is more than obvious. In 1874 Naples was the largest city of unified Italy. With its 450,000 inhabitants it was the fifth most populous city in Europe, and it had the greatest population density. Rent prices were high and the pressure on space extraordinary, which implied that throughout the nineteenth century most surviving green spaces were transformed into residential areas.[20] Moreover, Dohrn's station would be located in the Chiaia district, one of the busiest neighborhoods of Naples. It was built in the ornamental park of the Villa Reale that stretched along the coast. The street to the north side of the station was described in Baedeker's travel guide of 1880 as "the liveliest of the city." Especially on Sundays it was considered to be worth a visit, when one could see the spectacle of "four and even more rows of passing chariots, from the glittering equipage up to the droshky, the horse-car, the omnibus and the hurrying, mostly galloping double-wheeled 'Corricollo,' in an inextricable hustle." On the side of the bay, the sandy coast line would be replaced by a palm-lined embankment. The latter—again according to Baedeker—quickly turned into "the main stroll of Naples."[21] The large

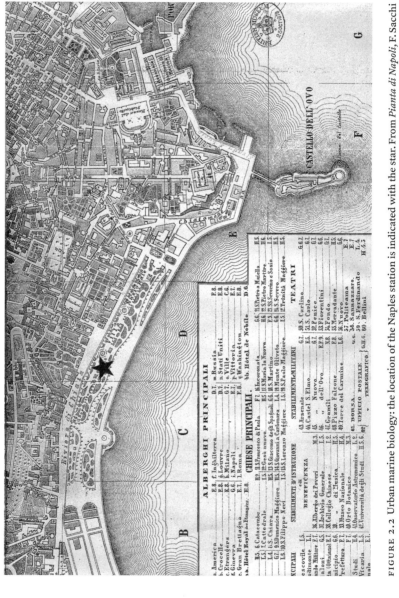

FIGURE 2.2 Urban marine biology: the location of the Naples station is indicated with the star. From *Pianta di Napoli*, F. Sacchi e Figli, 1877.

neoclassical building of the Stazione fit well in this environment. According to one early visitor, it was "often mistaken for a palace."[22] The cosmopolitan atmosphere was clearly an asset in drawing tourists, and it was also considered to allure vacationing scientists.

Not everything about Naples was appealing for visitors. To begin with, Dohrn himself would often complain about the region's climate. As early as 1876 he indicated in his correspondence that the temperature changes and heavy rainfall in winter and the heat and the mosquitoes in summer chased off visitors with weak nerves. The summer storms (called *sirocco*) often made boat travel dangerous and occasionally badly affected Dohrn's own mood.[23] More importantly, the climate brought malaria, while the general insalubrity and overpopulation made Naples vulnerable to typhus and cholera epidemics.[24] From the very beginning Dohrn feared that these could deter tourists and visiting scientists and harm the functioning of his station. On several occasions epidemics would indeed prove the Achilles' heel of the Stazione and Dohrn's personnel and family would also be personally affected.[25] The attractions of the city and the well-equipped station ensured, however, that both scientists and tourists always returned once the most pressing danger had passed. Next to the touristic charms and the laboratory equipment, scientists were drawn by the variety of species that were available in the Stazione. Indeed, Dohrn had chosen the site partially because of the diversity in animal life one encountered in the Naples bay. Furthermore, the Mediterranean species had a touch of exoticism for the northern zoologists who constituted his first target group.

The appeal of Naples might be clear. For observing animals in their natural environment, however, the place offered little prospects. As the hinterland of Dohrn's station was completely urban, inland excursions seemed pointless. Also the coastline itself contained little promise of successful observation. Excursioning marine zoologists in northern Europe had developed a tradition of "peering into the pretty tide-pools" to study animal habits, but this was impossible in Naples.[26] Not only had the original shoreline been replaced with an embankment, but the Mediterranean also has little or no tidal movement, which implies that in the wide vicinity of Naples tide pools were nonexistent. This left only the sea itself, which in the surroundings of the station was heavily polluted by the city's notoriously decrepit sewage system. Mussels and oysters caught in the bay would actually play a crucial role in the spread of epidemic diseases in the city.[27] The quality of the water was so bad that it had to be thoroughly filtered before it could be used in the aquariums of Dohrn's laboratory.[28] Naples did have a harbor, which made it easier to engage local fishermen to supply specimens from places in less polluted areas farther off. But for actually studying these animals in interaction with their environment, the Chiaia coastline was unpromising.[29]

Dohrn did not have a temperament to suffer passively the dictate of his station's geography. He wanted the Stazione to be an important center for *all* aspects of his Darwinian zoology and he tried to overcome the spatial problems he encountered. One way of doing this was to present the grand aquarium not only as a place for tourist entertainment, but also as a site for significant research. It might have been difficult to observe animals in the open sea, but this was no problem at all in the aquarium. All the zoologist had to do, Dohrn wrote in 1872, was "to sit in a movable chair, which is placed in front of the tank and which, thanks to special precautions, hides him and the tank completely from the general public." It then sufficed to place some rapacious fishes in the aquarium of, for example, the jellyfish to observe the struggle for life under one's very eyes.[30]

This was not an outlandish suggestion. Aquariums had after all been used for the study of animal behavior before. They had enabled Victor Coste, for example, to examine the nesting behavior of sticklebacks in the very center of Paris.[31] Dohrn seemed to be aiming in the same direction. He specifically engaged one of his assistants, the Austrian zoologist Richard Schmidtlein, for this kind of work.[32] The latter would set up long-time observations on aquarium-kept species, publishing his first conclusions in the station's *Mittheilungen* in 1879. There he would, among other topics, describe the mating behavior of stingrays and the defense strategies of crabs.[33] Yet this type of research was not really followed up by visiting researchers. In fact the Naples ground-floor aquarium mostly remained a place of bourgeois spectacle. Indeed, Schmidtlein himself seems to have played a role in this. In his visitor's guide to the aquarium, he described the life habits of animals as sources of "hilarity" rather than of serious scientific research.[34] Furthermore, he was known to contribute to the general atmosphere of amusement, by putting in the visitors' hands "their ancestor" *Amphioxus* and by having them touch electrical rays.[35] When Schmidtlein eventually left Naples because of health problems in 1881 the study of the *Lebensweise der Tiere* hardly seemed to have advanced. In a letter to Rudolf Virchow three years later, Dohrn recognized that all "energy and means" of the station had been absorbed by "pure morphological-embryological problems."[36]

A)

B)

FIGURE 2.3 Places of A, spectacle (the aquarium) and B, science (the laboratory) as represented in 1910 in *The Popular Science Monthly*. From Edwards, "The Zoological Station," 214 and 222 respectively.

He was eager, however, to put the study of animal behavior back on the agenda.

Dohrn's new initiatives concerned not armchair ecology, but rather the study of animal habits in the sea itself. By purchasing diving equipment he wanted to give the visiting zoologist direct access to the underwater world.

In 1878 Dohrn had already personally undertaken diving experiments in Kiel, together with the industrialist Werner von Siemens. The following year, he acquired a diving suit from the Italian Navy for his station. The fact that the Mediterranean had no tides was an advantage for diving and the clear water provided high visibility. Dohrn therefore was optimistic. In his letters to Germany he wrote of "an endless field of conquest" that lay before him.[37] Zoologists, so he stated, knew only about the life forms that could be seen from above the sea surface and those that were dredged up from the sea bottom. The area in between had been completely neglected.[38] Dohrn hoped to open up this field thanks to the diving suit. Similarly, he believed the diving equipment could help zoologists to access sea bottoms that were unreachable for dredges, like those in underwater grottoes.[39] Dohrn furthermore hoped to complement these underwater studies of animal life habits by breathing new life into the station's aquarium research. To du Bois-Reymond he wrote about his plans to create aquariums with movable walls. Animals attached to these could thus be taken out and subjected to all kinds of manipulations with regard to light, water pressure, temperature, and salt degree.[40] Dohrn saw limitless possibilities.

However, once again his projects met with limited success. The new type of aquarium seems not to have been realized, while the diving equipment did not really put visiting zoologists on the track of the study of animal habits. One visitor, the Belgian Julius MacLeod, commented that diving calls for a certain amount of experience and that it was not without dangers for the novice.[41] Like most visitors, he stayed for only a short time and did not go diving at all. Furthermore, those biologists who did take part in diving expeditions did not come up with many scientific results. All in all, such explorations were basically remembered for the great "emotions" and "curious souvenirs" they offered, not for their breakthroughs in the study of animal life habits.[42] Dohrn also recommended this special attraction to important guests and possible sponsors, who had no scientific objectives whatsoever.[43]

All this does not mean that Dohrn was unsuccessful in stimulating fieldwork altogether. In 1880, for instance, he successfully launched the monograph series *Fauna and Flora des Golfes von Neapel und der angrenzenden Meeres-Abschnitte*, which was set up to give a taxonomic overview of all animal and plant species in the Naples Gulf. The nicely illustrated books sold very well, not only to scientists but also to wealthy book collectors.[44] Yet, while Dohrn deemed taxonomy important, it was not the type of innovative fieldwork he had been thinking of in his original programs. In introducing the monograph series, Dohrn stressed that taxonomic studies were

FIGURE 2.4 An idealized "diver at work." In fact, setting up underwater animal behavior research proved a more difficult enterprise. From Nunn Whitman, "The Zoological Station" (1886), 793.

particularly meant to provide "naturalists, who turn to their laboratories for their special studies, a clear determination of their research object."[45] In this way his series was set up to offer a service to the study of morphology and physiology, rather than to ecological or behavioral studies. The latter part of Dohrn's research program, he admitted in 1886, had been taken up only to a very limited degree at the Stazione.[46]

Not easily dissuaded, Dohrn continued to look for technological stimuli in order to encourage research into the habits of animals and their interaction with the environment. In addition to an aquarium and diving equipment, he developed plans for a third locus of ecological research: the deck of a ship. In the 1870s, he had purchased two rather small steamships, but he felt that these did not have enough room to be used for scientific observations. The steamships sometimes did take scientists on board, but this was to sharpen their "spiritual power output" and to make them breathe some sea air.[47] Most of the time the ships were used only to collect specimens, preferably "as much and as quickly as possible."[48] In the mid-1880s, however, Dohrn was thinking of furnishing a third (bigger) vessel, which could be used as a "floating laboratory" and, thus, to enable research while at sea. This meant that the ship would actually accommodate the scientists and provide room and equipment for their research. Dohrn was thinking not only of zoologists; he also wanted to take archaeologists, philologists, and botanists on long and large-scale expeditions along the Mediterranean coast as part of a project to explore the region's human and natural diversity. Nonetheless, it was particularly for the study of animal habits that he thought his project promising.[49]

Dohrn believed that the floating laboratory would provide opportunities to do experiments on living animals in their natural surroundings. One of the experiments he suggested was a variation of his earlier aquarium plans. His idea was to let a stone slab down into the water and wait until all kinds of lower sea animals had become attached to it. After a while the stone was to be hoisted up and put in a basin on deck, where the animals in question could be studied. The experiment could be repeated as often as necessary and it was easy to play with variables such as depth, light, and inclination of the slab.[50] Like the urban lab the floating laboratory would offer the scientist a context of manipulation and control. At the same time its setting was one not of placelessness, but of a contingent and highly localized nature. The fact that one could repeat an experiment at different locations would exactly bring this local contingency to the light. Next to the epistemic benefits of a floating laboratory, Dohrn also saw practical ones.

He wrote to Virchow that an expedition ship would offer possibilities for the study of "questions of sea statistics and the economy of marine life." He was in all probability thinking about plankton studies and fisheries research as it was then being developed at the Zoological Institute in Kiel. Dohrn hoped to interest the Italian ministry for agriculture in this type of investigations and, by doing so, to tap new financial sources for his station.[51]

Yet, despite Dohrn's wild plans, the "floating laboratory" was not purchased. After an "inner conflict between the researcher and the family father," Dohrn decided that the acquisition of such a vessel would be too much of a risk.[52] As he had invested a great deal in the expansion of the station itself by opening a new physiology laboratory, there was no financial scope for his ambitious plans to perform experiments in the open sea. After the failure of the attempt to persuade zoologists to research animal habits in a diving suit or in a chair next to the aquarium, this actually meant the end of Dohrn's endeavors to stimulate the study of the *Lebensweise der Tiere*.

FACTORY EFFICIENCY

Despite Dohrn's inventive attempts, Naples proved a difficult place for launching animal life studies. This was to a large degree due to the physical landscape surrounding the Stazione. But also what Thomas Gieryn has described as the "normative landscape" played a role.[53] Places, Gieryn argues, carry their own norms about the desirability of certain behavior. In Naples indeed, immaterial values embodied in the managing style, social atmosphere, and self-representation of the station subtly interfered with the study of animal life habits.

To begin with there was the role of Dohrn himself. From the very beginning, he aimed for a large-scale enterprise. He wanted his station to be a worldwide reference point, and indeed most Western nations (except for France) were eventually to send large numbers of biologists to Naples. This magnitude called for the erection of an extensive edifice and Dohrn's continuing management. In the foundation years, he spent a great deal of time negotiating with the Naples city council, quarrelling with architects, soliciting his father for money, and networking with possible philanthropists and government officials. Once the station was established, Dohrn was constantly involved in fundraising activities, giving lectures throughout Europe and attending parties thrown by the rich and powerful. At the same time, he had to organize stays for considerable numbers of foreign researchers and dignitaries and manage a large group of staff, including fishermen, preparers, laboratory assistants, technicians, janitors, and a librarian.[54] His role as manager and fundraiser left little time to actually supervise the research that took place at his station. He might have wanted to stimulate the research program he had outlined in 1872, but in practice there were more mundane things to attend to.

Dohrn not only lacked the time to act as a scientific coach, the actual conception of his station also explicitly excluded this task. Dohrn promised his visitors total non-interference as regards research.[55] He did not guide the visiting biologists in their investigations and the studies done in his station had little or no influence on his own work, either.[56] This was a matter of money as much as of principle. It is clear, after all, that Dohrn was often frustrated with the concrete research projects that were set up at the Stazione. As we have seen, the neglect of research into animal life habits particularly disturbed him. Yet guiding his visitors into that direction seemed a complete impossibility. In a letter to du Bois-Reymond, Dohrn confessed to a feeling of powerlessness. He recounted: "Once I succeeded in directing an Italian to a remarkable undulating part of the dorsal fin of the *Motella* [or the rockling]—a piece of fin that moves as quickly as eyelashes— but to no avail, the foolish chap immediately discarded it to make a boring thirteen-in-a-dozen study on the epidermis of this fish. Most other men already arrive here with strict instructions. The only thing left for me is to wait until I can say: I buy your labor with 1500 Marks per year and you engage yourself to obey."[57] The actual organization of the Stazione, however, implied that it was not Dohrn who was to buy labor, but the governments and universities who rented research tables. As such, Dohrn's visitors would never be scientific lackeys.

Dohrn was never to act as a scientific master, and the profile of the average visitor to Naples further enhanced this. Unlike many other stations Naples did not receive zoology students or amateurs, but only experienced researchers. Dohrn's "table system" meant that biologists usually went through an official procedure of election by a scientific institution before receiving a place. Most of these visitors came only once or twice, and, more importantly, they came for fairly short periods. The majority of the biologists who visited the Stazione used academic holidays to carry out research that could lead to quick publication. Dohrn himself saw this as a major obstacle to the study of the *Lebensweise der Tiere*.[58] Such a research after all called for patience. Thus, biologists wishing to devote themselves to it preferably paid repeated visits to the site of study. This, however, seems not to have matched with the "hurry-up and publish mentality" that gained momentum under the biologists in the late nineteenth century.[59]

Dohrn might have deplored the mentality of some of his visitors, but he adapted to the fact that most of them came for only a few weeks. In fact, he actually buttressed the ideal of swift research and quick publication by constantly stressing the "efficiency" of the Stazione. From the foundation of his station "effective organization" played an important part in its self-presentation. In outward communication Dohrn at all times stressed that visiting biologists could find everything they needed close at hand: specimens, books, laboratory infrastructure, and well-trained assistants.[60] And it was indeed these elements that visitors hailed. In their letters home and in their reports in journals they talked of the "elaborate equipment," the "perfect appliances," the "order," and the "wonder of the specimens."[61] The station made it possible, so it was argued, "to accomplish the greatest possible amount of work in a given time, and with the least possible annoyance."[62]

In practice, Dohrn's stress on efficiency was to lead to the discouragement of field research. He defended a "division of labor," which implied among others that visiting biologists would no longer go into the field to procure their own specimens. Some of them regretted this development, but according to Dohrn his way of working was to the old organization what the modern railways were to the stage-coach.[63] In his laboratory, he argued, the zoologist no longer had to waste time acquiring his "material"; he could find it immediately on his working table. In short, the Naples station wanted to free the researcher from all the burdens of the old-style field expedition. Visiting zoologists did not have to drag heavy scientific equipment about, nor did they have to worry about their lack of familiarity with the country and the people or their inexperience with the technical aspects of marine zoology. They were able to start their laboratory work the very day they arrived and prepare a publication during a short stay.[64] In this manner, Dohrn argued, his station became a "time, money and energy saving power."[65]

The way Dohrn organized his laboratory actually widened the gap between the laboratory and the field, and thus discouraged visitors from studying organisms in interaction with their natural environment. In the Stazione, the specimens were brought in by "rugged Neapolitan fishermen" and selected by staff at the station, and only then were they delivered to the researcher.[66] Official regulations forbade visitors from procuring specimens directly from the fishermen. The researchers were allowed to go fishing themselves, but had to confer on this with the director first.[67] Efficiency, it seems, called for control. The spatial center of this control was the so-called sorting room. It was there that the specimens were brought by the fishermen and it was there that they were selected and divided among the visiting researchers. Visitors would often penetrate this "sanctuary" to gape at the new "surprises" that were brought in.[68] However, this was as close to the field as most of them got. Despite Dohrn's attempts to stimulate the study of animals in their natural surroundings, his rhetoric of efficiency seemed to undermine the very same project. It made the Mediterranean (in Dohrn's own words) into an "immense storehouse" of specimens, rather than a site for research.[69]

All this is not to say that the Naples staff lacked expertise about the life habits of animals. The leader of the sorting room, the in-house trained Italian Salvatore Lo Bianco, was rightly known for his "expert knowledge of the haunts and habits of every manner of marine creature."[70] Lo Bianco, for example, knew where to find various species of animals at several stages of their development. He did not, however, turn this expertise into a scientific project. He needed this knowledge in order to steer the fishermen to procure the material the scientists actually wanted. As such he deployed his expertise in a service of delivery of specimens, not in a research program on the *Lebensweise der Tiere*. Lo Bianco acted as the intermediary technician–"patting the fishermen on the back, talking seriously with the strictly scientific"—and in this way he was a crucial player in the efficient services of the station.[71] His services enabled many zoologists to work without ever coming into contact with the natural environment of the species they studied.

Lo Bianco's services of delivery were not limited to visitors to Naples. Curator of the station's Preservation Department since 1881, he also prepared specimens for shipment to be used in research, teaching, and exhibitions all over the world. He was known for his innovative techniques, which enabled him to preserve some types of invertebrates for the first time and to reach very "natural" results in most of his preparations.[72] His preparations might have been lifelike, but the animals were dead nonetheless—and they would be shipped to places far distanced from the habitat in which they lived.

The Stazione Zoologica looked like a palace, but it was run as a factory. This was a crucial part of its normative landscape. It involved large-scale ambitions, a managerial attitude of its leader, a division of labor, and an efficient handling of specimens as well as visitors. This approach certainly boosted particular types of research, but, as indicated, it also discouraged others.

Dohrn's double program called for both field and laboratory work. Embryology was indoor biology almost by definition, while the study of animal habits in interaction with their environment seemed at least incomplete without work in the field. In practice, the Stazione in Naples mainly offered visitors a laboratory environment and did little to bring zoologists closer to the field. At the same time it is important to stress that the lab and the field are not fixed entities, but rather evolving historical products. In Naples the laboratory was reinvented and to a large extent "naturalized." The Stazione certainly had its chair of descriptive stain-and-slice-research as was popular in the Germany of the 1870s and 1880s, but it also invested in experiments in addition to morphological description and in keeping living animals in addition to alcoholic collections. From the very beginning Dohrn prided himself on the fact that, thanks to his modern infrastructure, he was able to keep marine animals alive much longer than anywhere else.[73] Overall, he managed to use his station to make a world visible that had remained invisible for a long time. This natural world came to light in laboratory aquariums, under microscopes, and in Lo Bianco's jars with purified spirit.

Naples was a center of modern laboratory culture in biology. In fact, it played an important role in the rise of this culture itself, renewing and optimizing it. It contributed to the standardization of methods in experimental biology and became an international reference point as a site where (living or freshly killed) animals could be studied in "placeless" conditions. In sociological studies of the laboratory ample attention has been paid to both material and immaterial boundaries that keep the unauthorized out of such placeless labs.[74] The example of Naples shows how such boundaries could also work the other way around. In this chapter I have indicated how the walls of the sorting room in the Naples station, the physical constitution of the Naples bay, Dohrn's table system, and his rhetoric of efficiency all stimulated the authorized scientists to *stay in*. It may be clear that such an inward orientation was not simply rooted in an epistemological ideal of placelessness; it was also linked with both its physical and normative landscape.[75] Because of this double landscape a part of Dohrn's program remained unrealized. But, as the following chapters will show, similar programs would be taken up in different places, and this with quite different results.

CHAPTER THREE

Wimereux

Tide Pool Science

The year 1874 was not only the year in which Anton Dohrn officially opened the doors of his aquarium to the public. In the same year, some 900 miles to the north, the French zoologist Alfred Giard also inaugurated his marine station at the coast of the French department Pas-de-Calais. This event received substantially less coverage in the press, and the station in question was far less impressive. It was not a purpose-built palace-like laboratory, but a small rented chalet on the beach. Nonetheless it would become an important center of the rejuvenation of biology in France.

Giard started his work in Wimereux with a scientific program that resembled Dohrn's in many ways. Yet the science Giard and his visitors ended up doing was of a very different kind than the one performed in Naples. Whereas in the Stazione, the study of the animal's "habits and conditions of life" remained largely unexplored, it became one of the focal points of scientific activities in Wimereux. A closer look at these activities gives another instance of the power of place. Historian David Livingstone has stressed that in different localities "nature has been differently experienced, objects have been differently regarded, claims to knowledge have been adjudicated in different ways."[1] Marine stations illustrate this nicely. Compared with Naples, Wimereux offered different perspectives indeed.

The station of Wimereux presented the material and social preconditions that were needed to produce a new way of researching the animal in interaction with its environment. This novel kind of research was place-based ecology: it pursued general ecological knowledge through detailed studies of one particular site.[2] The approach would be central to the reform of fieldwork in the late-nineteenth-century life sciences as envisioned by several representatives of the station movement. It involved new kinds of observational and experimental practices and a new scientific language. Alfred Giard was one of the first to develop these, and Wimereux was the place where here did so.

ÉTHOLOGIE: THE PROGRAM

The program Giard eventually set up at Wimereux, and the means he deployed to materialize this program, clearly bore the marks of the training he had received in Paris. Both his vision of zoology and his conception of marine stations were heavily influenced by the work of his master, the celebrated Henri de Lacaze-Duthiers. The latter had been Giard's teacher at the Parisian École normale supérieur and he was the one who had introduced his student to marine zoology. In the late 1860s Giard undertook his doctoral research in Roscoff, at precisely the same time as Lacaze-Duthiers was setting up his station there. Like Lacaze-Duthiers and on the latter's suggestion, Giard started to work on the morphology of sea squirts. It was partially a dispute on the distribution of labor in the study of this subject that led to the eventual break between master and pupil. A quick appointment at the university of Lille, only two months after having defended his doctoral dissertation, put Giard in an independent position vis-à-vis his authoritarian former master. The foundation of his own station in Wimereux would be symbolic for this newly gained independence. In the following years, the rivalry of Giard with Lacaze-Duthiers only worsened. The first started to openly criticize the latter for being a one-sided anatomist, for his hesitance with regard to evolutionary theory, and for his overt anti-German sentiments. The resentment between the two men was particularly harsh and divided French zoology for decades. Yet Giard's intellectual patricide was not uncommon for ambitious nineteenth-century academics. Lacaze-Duthiers himself had broken with his mentor Henri Milne-Edwards in a similar fashion, and, as we have seen, Dohrn's clash with Gegenbaur and Haeckel had been particularly sour as well.[3]

The main bone of contention between Lacaze-Duthiers and Giard was evolution. Whereas the first pleaded for caution with regard to Darwinian theorizing, the second believed it should be the backbone of biological reform. As his coeval Dohrn, Giard particularly felt inspired by the Haeckelian school of evolutionary morphology, albeit that he (again as Dohrn) always aimed for a type of science that did not limit itself to the reconstruction of evolutionary trees. Giard believed zoology could be successful only if conceived broadly. For this reason he and his gradually growing cohort of students showed themselves highly critical of the increasing compartmentalization of science. The sharply drawn border between morphology and physiology (ever since the work of Claude Bernard) drew special criticism. Form and function were after all intimately linked, it was claimed. Giard therefore defended a synthetic program of the life sciences under the title of "experimental morphology."[4] Most of his pupils would opt for the banner of "general biology."[5]

In his broadly conceived program, Giard preserved a special place for what he called *éthologie*: "the science dealing with the habits of living beings and their relations, both with each other and with the cosmic environment."[6] This *éthologie* could count as the cement of his general enterprise. Analyzing the effects of the milieu on living beings would, after all, automatically involve both form *and* function. In this way, *éthologie* could tie morphology and physiology closely together and serve as the binding agent of "general biology."[7] It was in this context that Giard actively promoted the study of the interaction of animals with their environment. Between the 1870s and his death in 1908, he was able to convert a substantial number of students and friends to this cause.

The term *éthologie* might confuse present-day readers, since Giard's conception of it is closer to today's "ecology" than it is to "ethology," the contemporary term for the study of instinctive animal behavior. Even in Giard's own time the label *éthologie* seems hardly to have been used outside of his own scientific network. This does not mean, however, that he was working in isolation. Indeed the approach he was defending received momentum simultaneously in various countries under various names. Some terminological unpacking is therefore indispensable in order to situate Giard's project properly in the scientific context of its time.

The term *éthologie* was coined in 1854 by the Parisian naturalist Isidore Geoffroy Saint-Hilaire, who composed it from the Greek *ethos* (meaning "habit" or "habitual residence") and *logos* ("knowledge").[8] The expression, however, really caught on in France only after Giard began to use it in the 1870s. This success has to be seen against a wider backdrop. In the decades between 1870 and 1910, almost all Western countries witnessed a small but increasing number of naturalists pleading for the reappraisal of field biology. While these pleas lacked coordination, most shared the general idea that a "new" natural history had to be developed. This would be a new type of field science that would compensate for the excesses of late-nineteenth-century laboratory culture, while at the same time integrating some aspects of its methodology.

The new project was launched with various nuances and in various scientific contexts. Even when limited to zoology, the enterprise was polymorphous and the nomenclature confusing. Giard himself would refer to German *Oekologie* and British *bionomics* as foreign variants of his own French *éthologie*.[9] The term *Oekologie* was coined by Haeckel in his influential *Generelle Morphologie der Organismen* in 1866. It defined *Oekologie* as "the entirety of *the science of the relationship of the organism with the environment,* including in a broad sense all the conditions of existence."[10] Neither the term nor the program it described were an immediate success in the German states. Haeckel himself did little to develop the field and it was not until the late 1870s that the object of study was taken seriously by professional scientists—among others thanks to the efforts of men such as the Kiel museum curator Karl Möbius. The term *ecology,* however, only really came into use two decades later, and even then it was mostly botanists who associated themselves with it.[11] *Bionomics,* then, was used only by zoologists and was taken up in the Anglo-American rather than in the German context. The word first appeared in the *Encyclopaedia Britannica* in 1888 in an article by the English morphologist Edwin Ray Lankester. His term and the research field it described eventually gained a modest following in Britain and the United States around the turn of the century, but it never became a huge success. When, in the 1930s, the study of the interaction of animals with their natural environment eventually won popularity, it was under the name of *animal ecology.*[12]

Between the early French, German, and British research interests outlined above, there were some minor conceptual divergences. Yet the difference between *éthologie, Oekologie,* and *bionomics* was first and foremost one of scientific networks and personal contacts. In this sense the labels used referred as much to a scientific clan as they referred to a specific discipline. This was particularly the case for *éthologie.* All its important early propagandists were pupils or acquaintances of Giard, belonging to a limited number of academic circles in northern France and Belgium. In France, the most important were Jules Bonnier, Paul Hallez, Georges Bohn, Casimir Cépède, and Etienne Rabaud—who all studied with Giard at some point. The Belgians, including among others Paul Pelseneer, Jean Massart, and Louis Dollo, also met with Giard during their student days. Since the Belgian coast was considered ecologically uninteresting and for a long time did not have a functioning marine station, Belgian zoology professors tended to send their students to Wimereux. Several of them became part of the rather small clan that propagated Giard's *éthologie*.

Since it was a part of general biology, *éthologie* was not the only focal point around which Giard's group was organized. Giard also published widely on "pure" morphology and evolutionary theory, as did many of his followers. In fact, some of Giard's pupils entirely devoted themselves to other study objects.[13] For the outsider, however, it was clear that there would be no *éthologie* in France without Giard and his group. It was their teachings which provided the most important introductions to it, and it was Giard's *Bulletin Scientifique de la France et de la Belgique* which was its most important forum.[14] After Giard's death in 1908, it was his direct followers who kept it alive. When reconfirming the scientific program of the *Bulletin Scientifique* in 1909, they stressed that *éthologie* would continue to have a large place in the journal.[15] It was the same pupils who produced the most important publications on the subject until the 1930s.

Giard's network hardly crossed linguistic borders, but there were some isolated echoes of his work abroad. In Germany, Möbius's pupil Friedrich Dahl tried to breathe new life into his field of study by bringing it under the banner of *Ethologie*. Dahl used the French-inspired term because, in his view, Haeckel's notion of *Oekologie* was too narrow (because it excluded the study of the metabolism), while the old concept of *Biologie* was confusing (because it was also used to refer to the life sciences as a whole). But Dahl's French-inspired program never really caught on in the German scientific world. The Jesuit entomologist Erich Wasmann—one of the few naturalists to react to Dahl's pleas—argued that there was no reason to drop the term of *Biologie* for a new-fashioned foreign name.[16] In the United States then, it was the entomologist and taxonomist William Morton Wheeler who promoted the use of "ethology." His hope, as a zoologist, was that it would outcompete the ecology concept, which had been one-sidedly appropriated by botanists interested in plant geography. Yet also Wheeler's attempt was of little avail.[17]

In the longer term the label *ethology* survived but not without changing meaning. The German zookeeper Oskar Heinroth picked it up from the work of Dahl in the 1910s, and he largely linked it with his own special interest: the instincts of birds. It was his interpretation of ethology as the study of instinctive behavior that would, from the 1920s onward, be taken over by influential scholars such as Konrad Lorenz and Niko Tinbergen. Lorenz and Tinbergen's ethology was not to be organized around a loose network like Giard's, but would become rooted in a disciplinary infrastructure that survives until today. In this well-organized discipline the association with

Giard's clan—with its peculiar interests, methods, and concepts—was lost.[18] Giard's enterprise, as a consequence, was largely written out of history.

It is unhelpful, I believe, to present Giard's program of *éthologie* as a scientific stepping stone to classical ethology. Rather it was one of the more successful early forms of ecology, and one of the more efficacious attempts to set field science on a new footing. Overall Giard's approach was readily recognizable. His project of general biology might seem borderless, but his ecological research program was actually devoted to a limited number of species and was structured around a limited number of themes. With regard to species, Giard and his pupils largely focused on insects and marine invertebrates. Thematically, they concentrated on specific types of evolutionary adaptation. As such, the "Giardists" showed a sustained interest in topics such as mimicry (in which an organism gains an advantage by resembling organisms of a different species) and convergent evolution (in which different zoological groups have developed analogous characteristics through adaptation to a similar milieu).[19] The type of adaptation that intrigued the most, however, was that of the parasite. Both Giard and his pupils published enormously on various kinds of parasites and the relations they maintained with their hosts. As the host was literally presented as the habitat of the parasite, this subject matter was seen as prototypical for the type of ecology Giard pursued.[20]

It was indeed the study of parasitism that eventually led to Giard's most acclaimed theory—a theory that was exemplary for his approach. The study of a great number of invertebrates, ranging from insects to crustaceans, had convinced Giard that parasitism could strongly affect the gonads of the host organism, leading in some cases to sterility. The effects could be spectacular, since some animals took over the secondary sexual characteristics of the other sex. Giard coined the term *parasitary castration* to describe this parasite-related impairment of the host's procreation system.[21] It was a type of insight he was particularly proud of. Giard had at all times cherished the scientific ideal to discern general phenomena through numerous isolated observations. Furthermore, he was particularly eager to coin neologisms to describe these general phenomena—which, then, could serve as a marker of his scientific output.[22] Along with *parasitary castration* some of Giard's other neologisms had a continuing success in biology. One of these is *poecilogony*, a term he coined in 1891 to describe the varieties in embryonic development of animals of the same species that were confronted with different milieus.[23] Another is *anhydrobiosis*, an expression he launched in 1894 to designate "the slowing down of life phenomena under the influence of progressive dehydration."[24] All these examples are clearly the progeny of "general biology." They describe phenomena that have both morphological and physiological consequences and that could have been discovered only with the eye of the *éthologiste*.

The type of science Giard was after could not be carried out in all places. It needed a particular setting. The small fishing village of Wimereux, Giard believed, offered good prospects.

THE INNER FRONTIER OF PAS-DE-CALAIS

One of the reasons Giard selected Wimereux as the location for his station was because existing zoological catalogues showed a great variety of species along the rocky coasts of the region. Although in terms of species diversity Wimereux could not compete with Roscoff (let alone Naples), it was at least more appealing than the very uniform northern sand beaches in most of Pas-de-Calais and Belgium. Because of the presence of a small river, Wimereux was also considered interesting for the study of brackish water fauna.[25] Moreover, the surrounding habitats were diverse as well. Wimereux itself was situated in between cliffs, while further north one could find a large dune area, estuaries of small rivers, marine swamps, and a wooded valley with the Arcadian-sounding name "la vallée heureuse." Because of its relative density in water birds, the area exerted an attraction among hunters, some of whom traveled all the way from the United Kingdom.[26]

Like most coastal villages in the area, Wimereux was still very small in the 1870s. When Giard set up his station, the hamlet counted less than 200 inhabitants and the chalet he rented was one of the few constructions near the sea. On a normal weekday, no more than one or two carts would pass through the village's single street. In a contemporary British hunting story one could read that "a wilder or more retired spot it would be hard to wish for, and those persons who like to live *retiré de ce monde* (as the monks have it) could not do better than rent one of the Wimereux châlets."[27] The loneliness and isolation were precisely what attracted Giard to the place. He particularly considered the lack of a tourist industry an asset of Wimereux. In his inauguration speech of 1874 he stressed: "The absence of a seaside resort and the want of a luxury hotel makes Wimereux free from this loafing and unhealthy population whose idle curiosity is so annoying for researchers residing in seaports which are more fashionable and more renowned."[28] Science, Giard believed, needed retreat—as such linking up with an old cultural trope that associated knowledge with solitude. For ages, monasteries,

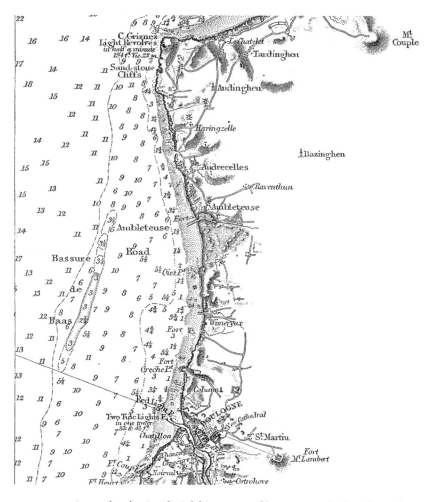

FIGURE 3.1 A map showing Pas-de-Calais's coast at midcentury. North of Boulogne there is only a series of hamlets in between the dunes. One of them was Wimereux. From *France: North Coast*, The Admiralty, 1858.

hermitages, or the wilderness had been presented as places were the individual could withdraw to find the truth.[29] Wimereux, unlike Naples, could be presented as such a place of withdrawal.

Yet the isolation and naturalness of Wimereux were relative. The English novelist who described Wimereux in 1868 as "a wretched wilderness" was more than overstating his case.[30] Rather it was one of the few remaining spots of more or less uncultivated land in a region with a quickly extending industry and intensive agriculture. Throughout the nineteenth century, monocultures of sugar beet and sunflower took over the landscape in large parts of Pas-de-Calais. The irrigated surface doubled, swamps were drained, rivers were canalized, and dunes were increasingly planted with pines.[31] A botanist publishing a flora of the regions Nord and Pas-de-Calais in 1878 indicated disappointedly that the plains of the area were "doubtlessly fertile for agriculture, but sterile and without interest for the botanist."[32] And it was not only the land that was increasingly exploited in northern France, but also the sea. In 1879, the port of Boulogne-sur-Mer—a mere five kilometers from Wimereux—employed no less than 4619 fishermen. Yearly they dredged up 20 million kilos of fish, one-ninth of the total national fish production.[33] While the fishery industry of Boulogne was having a growing impact on the marine ecosystem, the urban regions a few tens of kilometers to the north started what historians have described as "an orgy of production." The center of this orgy was the city of Lille. In the second half of the nineteenth century it grew into an unrivaled center for the textile industry and its population tripled from 75,000 in 1850 to 220,000 in 1901.[34] Hence the "wild" and "retired" Wimereux was in fact encircled by regions dominated by increasingly intensive agriculture, fisheries, industry, and urbanization.

Giard depended as much on this quickly modernizing world as he did on "unspoiled" nature. The appeal of his science lay partially in the promise that it would eventually offer guidelines for rationalizing the fisheries and agriculture—a claim we will further explore in the next chapter. The growth of Lille into an industrial center furthermore translated into rising scientific ambitions of the local administration. Lille's science faculty had been founded only in 1850, but by 1877 it already attracted the largest number of French science students after Paris.[35] Both the city council and the republican government invested in the university's scientific infrastructure. The politically left-leaning city furthermore offered a climate open to new-fashioned zoological ideas. Together with other rising industrial cities such as Nancy and Marseille, Lille was one of the few French cities that were relatively open-minded with regard to the theory of evolution.[36]

As a university professor Giard was of course by nature a city-dweller. It was therefore a major advantage of Wimereux that, thanks to a newly opened railway station in 1867, it was reachable in three hours from Lille. Researchers could come to Wimereux to collect specimens at low tide and study them in the university laboratory of Lille during the days that followed. The proximity of the university also made it possible to organize summer courses at sea. The botany courses would be taught in Lille on two days, while the rest of the week was devoted to the "practical teaching" of zoology in Wimereux. The hamlet's close proximity to Boulogne meant one did not have to bother about transporting provisions. In Giard's own words, Wimereux combined "the calm of the countryside and the contemplation necessary for serious study" with "the resources given by the proximity of a large city."[37] Wimereux constituted the type of place that Robert Kohler has described as "an inner frontier": a region that mixed relatively "untouched" nature with accessibility.[38] It is precisely for this combination that Giard selected it as the place for his station.

The scientific life along the inner frontier is obviously based on a shaky balance. When accessibility becomes too easy, the naturalness of an area may quickly erode. In Wimereux, the railroad would indeed bring not only zoologists, but also a rapidly increasing number of tourists. English holiday-makers and northern-French industrialists came to visit the recently opened race track and to stroll along the wooden sea wall. Around the turn of the century a casino was constructed and a series of grand hotels (with up to 150 rooms) opened their doors. The number of chalets rapidly multiplied. In Wimereux's high days more than thirty would be erected each year.[39] Giard found the rising tourist industry a nuisance, but did not immediately move out of his beloved premises. Even when he exchanged the university of Lille for the Sorbonne in 1889, he stuck to the makeshift laboratory in his rented chalet. Visiting zoologists nonetheless partially relocated their activity to the privately owned chalets of Giard's followers in the less crowded villages of Le Portel and Audresselles.[40] And when, in 1899, Giard eventually found the funding to set up an entirely new marine laboratory, he chose to construct it at Pointe-aux-Oies—a place considered more "natural," a few kilometers outside of the expanding village of Wimereux.

According to a visiting zoologist, Giard's new laboratory at the cliffs of Pointe-aux-Oies once more offered researchers a "relative isolation," far from "the indiscreet curiosity of bathers."[41] Again, however, the laboratory was not without its ties to the modern, industrial world. The land on which the laboratory construction was built had been offered to Giard by his friend Maurice Lonquéty, a local cement baron and a member of the French Association for the Advancement of Science. The same Lonquéty was involved in (eventually fruitless) plans to level out the dunes north of the new laboratory for the construction of yet another tourist resort.[42] It is clear that Giard, attempting to follow the shifting inner frontier of Pas-de-Calais, was compelled to forge some ambiguous coalitions.

B)

FIGURE 3.2 A and B, exploring the tide-pools in Wimereux, around 1920. (Courtesy of the Institut Pasteur, Paris)

Despite the annoyances brought by an expanding "civilization," Wimereux turned out to offer good potential for Giard's type of ecology. The large strip of land that became accessible at low tide made it possible to observe all kinds of marine animals easily in their natural environment. "Doing the tides" would become one of the crucial rituals of research in Wimereux. Specimens were of course taken back to the laboratory but, as one former pupil recalled, this was to check details of their behavior and physiology rather than to slice them up.[43] In addition to the beach, Giard would take his pupils and visitors on ecological excursions inland. Georges Bohn described the favorite walk of his former master, which combined different habitats—following cliffs, crossing rivers, and diving into the woods. Among other things it offered Giard the possibility of pointing out all kinds of episodes in the struggle for life.[44]

Sometimes Giard would organize boat trips on the open sea. As Wimereux did not have a harbor, this meant they had to use a vessel from the fishery station in Boulogne, which was directed by his former pupil Eugène Canu. Giard, however, approached these expeditions with limited enthusiasm, since the organisms dredged up were by definition removed from their natural milieu. Pupils were often excited to see the sea organisms "darting about in beakers" that they believed "had been created to live in alcohol," but Giard clearly preferred to visit places where animals could be observed in their natural habitat.[45] In one of his last articles, he wrote: "I attach no interest whatsoever in the capture of a species if I do not know exactly in which circumstances it can be found. The dredge and the sweep-net are, in my eyes, barbaric instruments that I only employ when I cannot use others and when the solution of an ecological problem seems impossible to attain by more intelligent procedures."[46] He therefore put a stress on slow foot excursions rather than dredging expeditions.

A ZOOLOGICAL HERMITAGE

The nature that scientists encountered in Wimereux was obviously of a different kind than the nature they found in the Naples Bay. But also the normative landscape of Giard's scientific chalet differed substantially from that of Dohrn's more famous marine station. This was partially a matter of scale. The small station at Wimereux basically had local ambitions and, though its appeal turned out to be international, the majority of its visitors came from northern France and neighboring Belgium.[47] Giard's chalet had room for only three visitors and, although there were often more than this, he never had to house the large number of scientists that Dohrn had to deal with.[48] Moreover, his staff had been kept to a minimum. Originally, Giard worked with only one personal assistant—who was an employee not of the station, but of the University of Lille. The limited staff implied there existed little specialization in tasks or division of labor. Looking back on this period, in 1900, a reporter of *La Nature* wrote that "the students, if not the professor himself, would take up the role of janitor, cleaning lady, fisherman, etc."[49] Because of the small numbers of visitors, there was not much managing to do in Wimereux, and this freed up time for Giard's own work and for focusing the attentions of his visitors on the research he deemed important. He spent a lot of time in the field with these visitors, stressing the value of ecological work in free nature.

In Wimereux, research tables were free and this enhanced the position of its director as a scientific guide. To be sure, Giard too would argue in favor of "full and absolute liberty" and some visiting biologists do seem to have carried out work that was outside of the scope of Giard's activities.[50] Much more than Dohrn, however, Giard tuned the studies of his guests to his own research interests. Former visitors might have stressed the lack of dogmatism and formal hierarchy in Wimereux, but they also presented the work done there as being a group enterprise—an enterprise which was clearly coordinated by Giard.[51] His organizing role in the station went so far that, in the early days, he would personally wake the visiting biologists in the morning to take them on field excursions.[52] In the last years of his life, he was seldom personally present among the visitors, but he trained his personnel to take over his role.[53] In the notebooks of his last preparer, Casimir Cépède, we find an almost endless series of observations preceded by the phrase *"Giard me signale"*—"Giard directs me to ..."[54] This attitude of the director was as important for the field orientation of the station as was its actual geographical setting.

The differences in managing style between Giard and Dohrn obviously relate to the differences in the populations of their respective stations. Whereas the station at Naples was designed from the outset as a research facility only, Wimereux was set up for the purpose of both instruction *and* research. These differences surely reflect the national peculiarities touched upon in the first chapter. While in France the marine stations developed in close relation with the universities, the German tradition led to independent institutes. So whereas Dohrn received only experienced researchers, Giard would also attract many of his own students (from Lille and later the Sorbonne) and apprentices from the École normale de Paris. Giard believed that young biologists should learn their work not in the auditorium on the basis of images or wax models, but by coming face to face with living nature. He was convinced that students had to look for their own specimens and that by searching, they would learn most. He stressed that he wanted to train "naturalists" and not "savants," like those he said he had met in Naples and who were not capable of doing their own collecting.[55]

As a teacher, Giard was omnipresent both on the excursions and in the laboratory, where he combined experimental pedagogy with a strong sense of brotherhood. Former students recalled how in the chalet everybody worked in the same room—and the master had no more space than anyone else. Giard himself said that his teaching could be summarized as "working under the eyes of his pupils" and many students would indeed mention this type of organization as the most instructive aspect of their stay at the station.[56] It might therefore come as no surprise that the "master" Giard influenced his visitors more powerfully with his visions of general biology than the "manager" Dohrn.

Historians have suggested (in the context of American biology) that it was especially junior biology students who were prepared to go on field trips, while the older, professional scientists would have preferred the "higher status" work in the laboratory.[57] In Wimereux, however, there was no such distinction. Giard actively tried to break with the idea that fieldwork was a low-status activity suited for only the young and inexperienced, and he did so with success. Many of his visitors and former students (including

Bohn, Cépède, Rabaud, Massart, and Pelseneer) were to continue their field-work as senior researchers—often using Wimereux as their operating base. Moreover, much of the work done by both Giard and his visitors referred expressly to the ecological experiences they gathered in the field.[58]

The shared field excursions in Wimereux strengthened the personal bonds and the regular visitors to the station quickly cultivated the idea of belonging to a "brotherhood" in which Giard was "a master as well as a comrade."[59] As such, Wimereux became a crucial place in fostering a scientific clan around Giard. It was indeed this clan that made up the backbone of *éthologie*, rather than well-established disciplinary structures. In this respect it was a typical product of French academic culture, in which personal networks around one patron played such an important role. These academic networks have been described in detail by sociologist Terry Clark. In late-nineteenth-century French academia, he argues, a typical network (or *cluster* in Clark's terminology) consisted of about a dozen people. It was mostly organized around one patron who held a chair at a prestigious university, but contained members who were active at various institutions. The members were often united by their own journal and (most importantly) by the patron's active intervention in the careers of the others. In all these elements, the network gathered around Giard did not differ from the rest.[60]

The marine laboratory in Wimereux was typical for late-nineteenth-century French scientific culture in the sense that it was connected with a patron rather than with an institution. So when, in 1889, Giard exchanged the University of Lille for the Sorbonne, he continued to be the director of the station. For his ecological research the station was far more important than his urban labs. In his laboratories in Lille and later Paris, Giard did not really create a community life, performing most of his own research at home. Only in Wimereux did he closely interact with pupils and visiting researchers. In later years his ex-pupils would often refer to the "family atmosphere" and the "sense of friendship" they encountered there.[61] In their commemorative practices—ranging from obituary notes and autobiographies to publications connected to institutional anniversaries—it was always the Wimereux station rather than Giard's urban laboratories that were surrounded with a sense of nostalgia.[62] *Éthologie* was a family affair.[63]

The family, to continue the metaphor, was furthermore an inclusive one. Besides students and visiting senior scientists, local amateurs also belonged to it. Among these, the most striking example is probably Alfred Bétencourt, a wealthy gentleman scientist, hunter, and collector who lived in the immediate neighborhood of Wimereux. Giard had met him while out collecting

FIGURE 3.3 The Giard clan in a crowded Wimereux chalet—from left to right: Alfred Giard, Paul Pelseneer (microscoping), Jules Bonnier (standing), Philippe François (reading). (Courtesy of François G. Schmitt)

specimens at low tide. The two men became friends and Bétencourt quickly integrated into the Wimereux circle. His elaborate personal collections eventually became a reference point for visitors to Wimereux, and they were often referenced in the station's *Bulletin*.[64] People like Bétencourt constituted an important contact point for Giard's circle with the world of amateurs, their local societies, and their culture of excursions.

Cépède's notebooks—the only substantial archive of Giard's circle that has been preserved—often refer to collecting trips in which professional scientists and amateurs intermingle.[65] These amateurs did not consist exclusively of traditional gentleman scientists for that matter, but also of a growing group of middle-class laypersons. It is in their societies that the idea grew of amateur work as a collective and rigorous enterprise, rather than the occupation of lone and aesthetically oriented gentleman scientists.[66] It is as such a collective workforce that Giard believed their work was important. In 1889 he wrote: "The harvest to be won is so rich that there will never be too many workers, and rather than show the amateurs the disdain that official science has too often shown them, we should strive to create in our country an audience of serious scientific dilettantes."[67] Again this attitude made Wimereux into a rather different place than Naples. In Dohrn's station the "table system" implied that workspace was assigned only to representatives of scientific institutions, thus excluding amateurs altogether.

In Wimereux the informal character of the station went together with a cultivation of its material simplicity. Giard and his pupils bemoaned the lack of an elaborate laboratory, but at the same time they took pride in mentioning the scientific results reached under these dire circumstances.[68] In the end, they were all to share a certain romantic view of the chalet, its community life, its uncurtained windows, and its poor infrastructure. Overall, the aura surrounding the station of Wimereux was one of asceticism. In this sense, too, Giard's station was clearly the opposite of Dohrn's. To indicate that in Wimereux no money was spent on lavishness, Bohn stressed that Giard had never aimed at founding a "gaudy aquarium"—implicitly distancing his station from the "palace" in Naples.[69] In the same vein, two other former visitors described the chalet as an "abbey" and its researchers as "monks," evoking an image of brotherhood, isolation, persistence, and sobriety.[70]

This type of rhetoric was not limited to Wimereux, but was typical for the late-nineteenth-century French marine stations in general. Lacaze-Duthiers, for example, made it a point to systematically warn scientific visitors that his station in Roscoff "was highly modest, lacking the splendour of Naples and other localities."[71] In the same vein, one of his pupils insisted that the architecture of Lacaze's laboratory in Banyuls showed "not a single molding, not a single exterior ornament," to add that the simplicity of its working laboratories echoed that of "monastery cells."[72] Such assertions obviously matched well with the ascetic positivism in vogue among Third Republic ideologues. In this ideological atmosphere, efficiency did not particularly matter. In his articles, Giard rather stressed the importance of endurance for the naturalist. The ideal visitor he had in mind for his station was a patient, unwearied observer who did not mind coping with the problems and contingencies of the field, while he abhorred the spoiled indoor "microtomist" looking for an easy publication.[73]

Giard's example shows that the lab-field border was not only a place where laboratory ideals moved into the field. It was also a place where field values were integrated in laboratory practice. Giard's stress on the importance of isolation, tenacity, the ability to manage for oneself, and the willingness to "suffer for science" reflected ideals that had traditionally been connected to field exploration rather than laboratory work. Rebecca Herzig has compellingly indicated how, throughout the late nineteenth century, "the physically and emotionally taxing work of [field] exploration" gradually started to emblematize science as a whole.[74] In Wimereux the same thing happened. The values deemed important during excursions followed the Wimereux regulars once they entered the chalet-laboratory.[75]

In the 1890s Giard would move his marine laboratory to a new building. The new architecture, however, stayed faithful to the old rhetoric and self-image. The new laboratory was bigger than the original chalet, but still rather small in scale. It had some sleeping quarters for guests and a series of work tables, but its design was far from palace-like. The cottage style of the building gave it a certain rusticity, which differed greatly from the more urban neoclassicism of the station in Naples.[76] Giard's new station was erected a few kilometers outside Wimereux, which in the preceding years had slowly turned into a bathing resort. So when, in 1905, the ornithologist Auguste Ménégaux published a detailed description of the station, he was able to follow the clichés Giard and his pupils had used in their earlier narratives on the life in the chalet. He stressed that the station was "far from luxurious," that it was located "at a distance from the frivolous activity of the well-known beaches," and that it was a place of "rest and work" and of "good and healthy camaraderie." In short Wimereux offered "a zoological hermitage." The accompanying illustrations show the station isolated amid nature.[77] The angle of the photographs is chosen to hide the neighboring Hôtel Cosmopolite.[78] Wimereux, so it seemed, was far from cosmopolitan. It was presented as a lonely outpost in the field.

NATURE'S EXPERIMENTS

Place was important to Giard's science. And Wimereux, rather than Lille or Paris, exemplified this importance. Giard and his pupils took care in describing the Wimereux landscape, the architecture of the marine station, its (lack of) infrastructure, and its social atmosphere because these elements carried strategic importance for their scientific project. It was all part of launching a new way of approaching the natural world. The Giardists propagated a study of "free nature" that was more than simply descriptive. They hoped to create a science of the interaction of animals with their natural environment that was as "explicative" as studies carried out in artificial laboratory conditions.

The new enterprise of explicative field science was presented as the rejuvenation of an old tradition of natural history—of which Giard deplored the decline in popularity.[79] His type of ecology, so he indicated, continued the interests of this tradition, embodied by eighteenth-century naturalists such as René-Antoine Ferchault de Réaumur and Charles de Geer.[80] Since the eighteenth century, Giard claimed, the work of field naturalists had not become less important—despite the fact that urban scientists had started to mock them as "insect-pinners" or "inoffensive monomanes."[81] He proclaimed that even when these naturalists performed purely descriptive systematics they lay important groundwork for more interpretative science. At the same time, he made clear that the ambitions of his own group were not in the field of insect-pinning. The appeal of inventories and taxonomic labels, so he stressed, was "a stadium of intellectual culture that can be nothing but transitory."[82] In order to reach a higher intellectual level, then, it was necessary to promote the field as a place one visited to study nature in action, rather than to kill and collect.

Giard's project had a negative and a constructive side. Before expounding on which science he endorsed, he took a lot of time to indicate the tendencies of modern biology he abhorred. One of these tendencies was an infatuation with technology. In the words of one of his students, Giard combined his field ethos with "a sovereign disdain for technique."[83] Again and again he stressed that studying nature *in* nature would offer great insights "without complicated apparatus, without expensive installations and without priggish *mise en scène*."[84] In Wimereux, the lack of a good laboratory infrastructure seemed to drive researchers outside, but Giard was happy to see this in a positive light. And the visitors to the Wimereux station followed his reasoning. In their letters, they often deplored the "deficiencies of the installation," only to add that (under Giard's guidance) they were still able to learn a great deal outdoors.[85] The most important insights, after all, were thought to reside in the "vast laboratory of nature," rather than in the modest "real" laboratory at their disposal.[86]

Giard combined this type of rhetoric with an open attack on what he perceived as the exaggerated laboratory orientation of his time. He argued that the technical advance of the modern laboratories had bred many "cabinet naturalists" and "alcoholic dissectors" who were out of touch with "living nature." In Germany in particular, he argued, several zoologists had become too fascinated by the slices of dead tissue provided by their microtomes. Giard was repeatedly to chastise the modern "sect of microtomists" who, in his view, represented a completely one-sided zoology. In 1878 he wrote of the microtome vogue:

> One hardens [the tissue], one colors it, one gives it the right consistency in a convenient mix; and then: it's time for the machine! Slices follow each other; one draws everything which is a little bit peculiar; one describes it with convenient care, and all this fills a memoir. Does that make you into a zoologist? I answer you without hesitation: *no*.[87]

With such remarks, the Giardists countered the idea that there existed an indispensable epistemic connection between experimental science and the laboratory. Much laboratory work, so they claimed, was not experimental at all. Many histologists and microtomists, after all, produced descriptive work of which the only use was "to fill the numerous periodicals [...] and embarrass the bibliography."[88] Giard's pupils would elaborate on this criticism and give it an echo in the early decades of the twentieth century. Their praise of a renewed natural history went together with an explicit denigration of laboratory workers who were interested only in "dead tissue." This denigration stands out clearly in the work of the Belgian botanist and Wimereux regular Jean Massart. With regard to the modern laboratory biologist, he wrote:

> Woe betide the plant or animal which they spot on their field expeditions, because their only interest is to capture it and to bring it to the laboratory, where it will be cut into pieces. A plant can offer the most eccentric adaptations, the habits of an animal can be as strange as possible: all that does not matter; it is "material" that they are looking for, nothing more; they only come to the countryside to raid.

Massart did not write off the laboratory as such—he even called himself a "laboratory man"—but he stressed that biology would always be incomplete without "ecological observations in nature itself."[89] The same could be heard among the other members of Giard's group.[90]

Despite this returning mockery, it was not so much histologists and microtomists who constituted an epistemic threat for Giard's enterprise. Like many of his fellow zoologists, Giard believed it was rather Bernardian physiology that represented the main peril. The physiologists, he feared, were going to monopolize the aura of innovation in the life sciences. This was to lead to a science that was narrow in many respects: in terms of location (urban laboratories), in terms of animals studied (dogs, rabbits, frogs, and guinea pigs), and in terms of appliances used to study them (syringes and recording cylinders).[91] The main reproach Giard addressed to the physiologists concerned their definition of experimentalism. As we have seen in chapter I, this reproach echoed earlier criticism by Lacaze-Duthiers and Victor Coste. In line with this older generation, Giard and his followers would stress that an experiment (*expérience*) did not necessarily entail the manipulation of nature. Like Lacaze-Duthiers, they stated that an experiment simply recorded causal relations. This implied that in order to have an experiment a hypothesis was to be followed by a series of observations. In some cases, the observation might even precede the hypothesis of an experiment, so that it was "only the spirit of the observer that gives it this [experimental] character."[92] Observations in free nature could in this way receive an experimental status, if, at least, they registered the causality of events.

Giard, of course, realized that free nature was unpredictable, and that his "experimentalism" depended partially on coincidental observations. Whereas the environment in the laboratory was rather easy to control, this was obviously less the case for the field. Giard admitted chance would always play a role in his scientific practices, but tried to limit its importance. One way to do so was to limit one's attention to one particular place—in this case Wimereux. Familiarity with the site would, after all, make it easier to discern cause and effect. This necessitated long-term presence. One of the main rationales of the marine station was exactly that it could offer this. Early-nineteenth-century expeditions offered only short temporal slices of nature, but the station promised permanent monitoring.[93] In this way, the Giardists claimed, their work could be distinguished from typical lab science. Wimereux regular Paul Pelseneer, for example, indicated that many artificial laboratory experiments were "childish" because of their "minimal duration."[94]

As part of their long-term monitoring, the Giardists tried to mobilize as many eyes as possible. For this reason they called upon lay persons. Unlike most laboratory workers, whose networks were largely confined to their own professional group, the Wimereux scientists drew many nonprofessionals into their enterprise. In Cépède's field-notes one finds reference to observations of fishermen, game wardens, and lighthouse-keepers, for example. In his notebooks, Cépède even recorded the conventional wisdom of villagers as indications about when and where he could encounter certain species.[95] The references to these people often expired in the filtering

FIGURE 3.4 An impression of the Wimereux chalet by Louis Bonnier—artist and brother of zoologist Jules Bonnier. From Giard, *Oeuvres diverses* (1913), 13.

process of scientific publishing, but some allusions to their folk biology re-mained. Giard, for example, regularly referred to fishers' terminology and recorded some of their myths with regard to the natural world.[96] On other occasions he explicitly mentioned the help he received from farmers or schoolteachers.[97] Yet, at the same time he made sure to clearly distinguish his own science from their practical knowledge.[98]

Familiarity with a certain place and access to a group of local fieldworkers were only a starting point for developing an experimental field science. In the field favorable chances to perform "experimental" observations were believed to be few, and therefore the observer needed a "training of the eye and perspicacity that one only acquires with serious efforts."[99] Rather than surveying at random, the naturalist needed to approach nature with hypotheses at hand. Observation, Giard claimed, is more intense and fruitful when one knows what one is looking for and when the gaze is theory-driven. He saw his own observations on parasitary castration as a good example. Once Giard had developed a hypothesis about the sexual effect of parasites on their hosts, he started to check his theory on the barnacles, brittle stars, leafhoppers, and daisies he encountered in Wimereux. Nature might have prepared the experiment, as Giard would phrase it, but he had to cut open a lot of organisms to actually witness it.[100]

Giard's activities needed so much legitimation because they ran counter to the growing consensus of what observations and experiments actually were. Lorraine Daston and Elisabeth Lunbeck have described how the
nineteenth century witnessed a growing "methodological opposition of active, reasoning experiment to passive, registering observation."[101] By mixing up the categories, Giard conceived a science that was atypical for his
time. Obviously, Giard and his pupils were well aware that nature's experiments were structurally different from experiments in the lab. In a typical
laboratory experiment, the scientist manipulates one or two variables, and
then compares the situation before the manipulation with the situation after. Nature's experiments lacked the material intervention of the scientist,
but worked with similar comparisons. These comparisons were often of a
temporal kind, assessing the situation before and after a certain change in
nature. This change could be induced by natural agents, such as a parasite
changing the morphology of its host. But, the Giardists were equally interested in (passively) observing the changes induced by human activity exterior to their own. The fact that the Belgian Yser region had been flooded
during the First World War gave Massart an ideal chance to study the botanical recolonization of the area "experimentally" in the postwar years.[102]

Such time-based comparisons were only one way to conceive of natural
experiments. More often the Giardists would present spatial comparisons
as the most fruitful approach to unpack the experiments of nature. This
involved comparing the well-known situation in Wimereux with those in
other sites. To make this possible, Giard set up several excursions. For his
study of the parasitic crustaceans Bopyridae, for instance, he worked a while
in the station of Concarneau (with the leaders of which he had close relations) and he stayed in coastal towns such as Fécamp and Pouliguen. He
even went to Roscoff, but, obviously, without visiting the well-equipped station of the competing clan led by Lacaze-Duthiers. The various species of
Bopyridae Giard encountered turned out to be numerous at some places,
and completely lacking at others. Giard traced their restricted habitats and
researched the conditions that determined their spread.[103] At other times,
the comparative method was exploited further and led the Giardists to compare the functioning of similar species living in different environments. The
most elaborate example of this approach is to be found in the doctoral thesis
of Giard's Sorbonne assistant, Georges Bohn, which compared the respira-
tory habits of crustaceans at six different places on the French coast.[104] The
converse strategy was also used, namely comparing the functioning of *different* species of organisms living in *similar* habitats. A good illustration of

this type of approach can be found in the dissertation of Bohn's friend, Casimir Cépède, who indicated how the common parasitic lifestyle of very differing species of protists had led to striking similarities in their morphology.[105]

Comparison, either temporal or spatial, was thus a crucial method in making field science experimental. This conflation of the comparative and the experimental was nothing new of Giard's or even Lacaze-Duthiers' generation, but can be traced back to the early-nineteenth-century work of Georges Cuvier. In *Le règne animal* (1817), the latter defended comparison as "the most effectual mode of observing [...] [the] experiments ready prepared by Nature."[106] At the same time, however, Cuvier believed that, in practice, comparison was best performed from one's anatomical cabinet, where specimens were necessarily detached from their natural habitat. His comparisons notoriously concerned the internal structures of the organisms under consideration, without paying attention to their wider living environment.[107] Giard, to the contrary, believed comparison made sense only when the organism's habitat took center stage. A good example is his study on the eggs of the shrimp *Palaemonetes varians*. In this case, the "natural experiment" consisted of comparing the number and size of the eggs as they were found in Wimereux and in freshwater lakes surrounding Naples. Because the Italian eggs were bigger in size and smaller in number, Giard concluded that the struggle for life was less harsh in freshwater and that the Italian shrimps could therefore invest more energy in nutrients.[108] Many of his followers set up very similar "experimental" comparisons that combined Cuvier's preferred comparative method with a strong sensitivity for the local particularities in the field.

Other types of "experiments" were closer to the laboratory model, in the sense that they involved more manipulation. A good example is Giard's study of the common fleabane—one of his botanical detours. On the road of Wimereux, he encountered some teratological forms of this uncommon plant. To study the possible spread and heredity of the morphological abnormality, Giard did not take the plant to the laboratory, but instead he removed all the normal variants he encountered. As such, he could study heredity "in the wild."[109] Massart, similarly, would move inland plants to the coast and vice versa, to observe the effects of this dislocation on the struggle for life.[110]

Whether conceptually or materially, the work of the Giardists often involved manipulating the places in which they worked. As such, they would become experts in what Robert Kohler has called "practices of place." These practices offered a way to turn observation and comparison into tools to

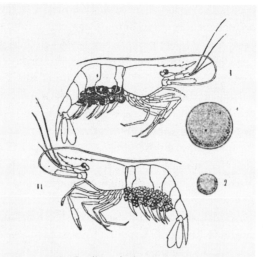

Fig. 12.— La *Pœcilogonie* chez *Palœmonetes varians*.
I, femelle adulte d'eau douce (Naples). — II, femelle
adulte d'eau saumâtre (Wimereux). 1, œuf du type
d'eau douce de grandeur comparative à 2, œuf du
type saumâtre.

FIGURE 3.5 The impact of the environment: Shrimps and their eggs from (top) Naples and
(bottom) Wimereux. From Giard, "De l'influence" (1911 [1889]), 398.

research causal relations.[111] The concept of nature's experiments was crucial in that regard. Yet, even with nature conceptualized as a laboratory, the
Giardists often had to fall back on a "real" laboratory—for a microscopic
study of a parasite or of the embryonic development of a crustacean, for
example. At the same time, the field ethos accompanied them when entering the lab and made them undertake various attempts to "naturalize" the
laboratory environment in which they worked. One way to do so was to
keep animals alive and breed them, rather than to dissect them immediately.
In the original Wimereux station this was quite an undertaking, as it was
underequipped and did not even have running water. Giard nevertheless
managed to keep all kinds of creatures alive in crystallization bowls and
primitive aquaria for at least a few days.[112] Especially aquaria could provide
pseudonatural conditions to study phenomena that were difficult to observe
in the wild. Bohn, for example, set up aquarium experiments to analyze
the reaction of marine flatworms when confronted with waves.[113] Aquaria
could also be useful to verify hypotheses that were developed in the field.

Cépède, for instance, believed on the basis of field observations that drought in swamps led to the metamorphosis of dragonfly nymphs, followed by their immediate migration. To authenticate this, the drought scenario was repeated in an aquarium in the laboratory.[114] The Giardists continually traveled back and forth between their lab and the surrounding field.

By the turn of the century, both Giard and some of his followers moved more and more in the direction of strictly laboratory-oriented research. It was particularly the physiological work that the German-American Jacques Loeb devoted to tropisms (or the "forced movements" of organisms triggered by stimuli from the environment) that inspired them for this move.[115] Following Loeb's example they set up experiments in which they analyzed the effects of light, temperature, and the salinity of the water on the reactions of all kinds of invertebrates. In this way, "local" knowledge apparently became less important. At the same time, Giard and his pupils still observed their organisms with an eye trained in the field. Bohn, for instance, would stress that even when one studied animals in a laboratory setting it was of importance to know its antecedents in the wild. Laboratory circumstances should, after all, perfectly mimic those in "real" nature.[116] Furthermore he argued that, unlike experiments in physics and chemistry, it did not suffice to have equal circumstances at the moment of testing when performing a biological experiment. Earlier environmental influences could have a long-lasting influence on the animals that were used. Therefore it was crucial to put the organisms in equal conditions long before the actual experiment was performed.[117] In other words, Bohn's eye for the importance of "local" conditions made him extra vigilant when trying to set up experiments with "universal" ambitions. His attitude had accompanied him from the beaches and the tide pools of the French coast into the secure walls of his Paris laboratory.

The rhetoric of Giard and his pupils is revelatory with regard to the state of the life sciences in turn-of-the-century France. The Giardists felt the need to explicitly and elaborately defend the value of their observations in "free nature." At the same time their work involved a lot of indoor microscoping, cutting, manipulating test tubes and acids, but these activities did not seem to need further legitimization. This is related to the almost self-evident prestige laboratory work had acquired in the late nineteenth century. It is also related to the strategic self-presentation of the Giardists. Although their science encompassed continual border crossing between the laboratory and the field, they particularly associated themselves with outdoor science. In this way they could present themselves as a countermovement in a period of one-sided laboratory research. It was a matter of identity building and boundary work. It involved ridiculing microtomists and physiologists, and redefining what an experiment actually is.

The Giardists were taking part in a struggle over where "nature" resided. This explains their explicit concern with the spatial aspects of their science. The marine station of Wimereux and its surroundings constituted only one of the places where they performed research, but they granted it a privileged status. It was there—far from the Parisian physiology laboratories—that they came closest to real nature. For this reason the site of the station was carefully selected and its perceived isolation was elaborately associated with virtues such as patience and endurance. Giard and his pupils developed a series of spatial strategies to turn the station into a center of "explicative" science. These strategies made the Wimereux landscape into a reference point for cross-geographical comparison and a preferred place to witness "natural experiments." Furthermore, these experiments served as the model for those that would be performed in the laboratory.

Giard's scientific program was one of broad biology not all too different from the one pursued in the more renowned Stazione of Naples. Yet, its outcome was very unalike. This has to be explained by the fact that, unlike the Stazione, the Wimereux station could—both practically and rhetorically— act as a workplace *in* nature. This was enabled by its location in between railroads and tide pools, by its moral economy of patience and austerity, and by the conceptual framework of broadly interpreted experimentalism. Urban laboratories might offer the scientist a research object that was easier to access, to control, and to interpret, but according to the Giardists this object would always remain an incomplete re-enactment of real nature. Compared with the experiments of nature, laboratory experiments could be only piecemeal imitations.

Studying the relation between organisms and their environment, so this story shows, was not just a matter of going out and making observations. It was also a matter of creating the material, moral, and epistemic conditions in which these observations could take place. In Giard's case, it was a chalet on the beach that was crucial for generating these conditions.

CHAPTER FOUR

From Wimereux to the Republic

Individuals and Their Environment

One small coastal village had formed the backdrop of Alfred Giard's type of ecology. It was in the dunes of Wimereux, along its cliffs and tide pools and in a chalet situated on its beach, that many of the concepts and practices of Giard's ecology originated. He favored a science that took local circumstances seriously, thus forcing researchers to thoroughly familiarize themselves with one particular place. It was the local world of shrimps, crabs and their parasites, of fleabanes, dragonflies and flatworms that took center stage.

The marine station of Wimereux was the place where Giard developed a particular way of looking and where his scientific network of pupils and followers took shape. This network would eventually help in spreading Giard's ideas through the French-speaking world. In this process, Giard's ecology would be applied to new objects and integrated in new working contexts. The spread of his scientific approach went in several directions. Developed to study "free nature," it would be applied to the world of caged animals. Initially focusing on the study of the living world, it would be adapted to examine long-extinct prehistoric species. At first concentrating on undomesticated regions, it would be used for understanding strongly cultivated agricultural lands and overfished seas. And, although originally focusing on the lives of invertebrates, it would be transposed to develop insights in human societies.

The story of the spread of Giard's scientific approach is not one of discipline building. Giard's *éthologie*, after all, would never possess the paraphernalia we generally associate with a proper discipline. No university chairs would be established that were explicitly devoted to it, nor would there be journals or conferences. It introduced a new scientific attitude rather than a clearly bordered branch of learning. It was associated with the particular interests of Giard himself, and its spread largely coincided with the growth of his scientific clan. In the work of the members of this clan a new set of questions was taken to be relevant, new kinds of observational and experimental practices came to the forefront, and a new language was developed. This chapter focuses on how, by the use of Giard's personal network, this new attitude traveled from one discipline to the next. *Éthologie*, indeed, proved to be a transdisciplinary undertaking par excellence. It was obviously of importance for zoology and botany, but it also found its way to paleontology and museology, agronomy, applied fishery research, and sociology. Spatially, it was largely associated with the field (which the ecologists re-evaluated as a place of scientific inquiry), but it also became connected to the laboratory and the museum. And it inspired at least one would-be zoo designer. As a scientific attitude, *éthologie* was clearly flexible enough to be adapted to different spatial and disciplinary settings.

The flexibility of Giard's program is not the only explanation for its success. With its interest for the interaction between individual organisms and their environment, the work performed in Wimereux tapped into a wider fascination of the late-nineteenth-century French and Belgian bourgeoisie. In this period several leading politicians and intellectuals adhered to environmental determinism, and with it to the conviction that individuals (both animals and men) could be changed by changing their environment. Giard's projects were both shaped and sustained by the cultural success of this conviction. This at least partially explains why his work received political backing and intellectual acclaim.

This chapter leaves the world of marine stations, to explore how a scientific approach that originated there could receive wider resonance in the science and culture of its time. Giard might not have created a discipline, but many researchers started to look at the world with the "trained eye" he had been propagating. This chapter traces the spread of his way of looking. To do so, it first focuses on the intellectual and political milieu in which it was able to flourish. Second, it looks at its translations in, subsequently, science popularization, applied fishery and agricultural research, psychology, sociology, paleontology, and zoo keeping.

PARIS, ANZIN, AND REPUBLICAN POLITICS

When Giard and his pupils studied protozoans, crustaceans, or insects, they never looked at these organisms in isolation. It was the effect of the environment on their form, functioning, and habits they were interested in. How did these animals respond to parasitism, dehydration, or a changing salinity of the water? Such responses to the environment were deemed important, not only for the immediate changes they brought in individual animals, but also for understanding the long-term evolutionary process. In all the publications of Giard and his pupils, ecology was closely connected to evolutionary theory. Echoes of Darwin's ideas were prominent in their work, but Giard and many of his followers also incorporated the French tradition of Lamarckism.[1] In this way, their methodological stress on the importance of the environment went together with the belief that individual adaptation to the milieu played a crucial role in evolution. At least some of the changes triggered by the environment were considered to be inherited by the offspring of the organisms in question (the so-called inheritance of acquired characteristics). And all this was not just a matter of science: it was heavily intertwined with politics. Giard was a member of parliament for the radical republican party in the 1880s and many of his pupils were on the reformist left as well.[2] Whether "radical," "solidarist," "progressivist," "socialist," or "left-wing liberal," their political convictions showed too much convergence with their scientific ideas to be coincidental. In both politics and science most of the Giardists strongly believed in the perfectibility of living beings, whose (Lamarckian) adaptations were the motor of progressive evolution.[3]

The associations connected to Giard's ecology were these of Lamarckian evolutionism, but also of mechanicism, anticlericalism, and radical politics. These connotations could obviously hamper Giard's career, but at certain moments the opposite was equally true. Particularly after the take-over of the political left in the late 1870s, the intellectual climate in France seemed receptive for the type of science Giard was propagating. In an atmosphere of republican anticlericalism and scientism, the Giardists had opportunities to thrive.[4] Giard's appointment to the Chair for the Evolution of Organized Beings in Paris in 1888 is a case in point. The chair had been created on the initiative of Léon Donnat, a liberal-radical councilor of Paris. The latter was a mine engineer, positivist, and author of a book with the programmatic title *La politique expérimentale* [Experimental politics] (1885). With the new chair (originally labeled "biological philosophy") Donnat hoped to set out a drastic reform of French philosophical teaching, detaching it from metaphysics and tying it with progressive science. By acknowledging evolutionary theory, the chair was furthermore believed to help France in intellectually catching up with its neighbors. It was explicitly designed to link up again with the high days of Lamarck—around whom a republican personality cult began to crystallize.[5] In this context the choice for Giard as first holder of the chair might not surprise. Both his republican and Lamarckian credentials were clear. Furthermore he had successfully staged himself as a victim, whose career had been blocked by a purportedly conservative and authoritarian Parisian scientific elite. Donnat and his ideological allies were determined to infiltrate this elite and Giard was happy to be their man.[6] His appointment at the Sorbonne was more than an honor; it enabled Giard to strengthen his position in the French zoological landscape and to launch the careers of his pupils. Giard's ecology might have been performed most easily in Wimereux, but he had to be in Paris in order to successfully propagate it. It was there, after all, that political support could be won.

For the many Belgian pupils of Giard the situation was somewhat different. After the Catholic Party took power in 1884 it would dominate Belgian politics for thirty years, which implied there was little official support for evolutionary theorizing. This was not without consequences. Paul Pelseneer, for example, believed his career prospects at the State Universities of Ghent and Liège were thwarted by direct Catholic intervention.[7] Yet, despite Catholic dominance, important liberal and positivist strongholds remained, and these were generally more sympathetic toward a mix of biological thinking and left-wing politics. The most important of these strongholds was without a doubt the University of Brussels.[8] It is no coincidence that of all Belgian universities, Brussels would send the most students to Wimereux. Jean Massart and Louis Dollo were offered professorships in Brussels; Pelseneer, eventually, an honorary doctorate.[9]

Giardist ecology received support in particular political circles only because it was believed to contain ideological consequences. This does not mean, however, that the Giardists necessarily drew explicit political conclusions from their field research. In the writings of Giard, the link between biology and politics was largely an unarticulated one. Despite his combined scientific and political career, he was generally cautious to keep the discourse of both spheres separated. If he touched upon the ideological implication of ecological notions, it was mostly to reprimand simplifying metaphors.[10] Others showed less caution. A friend of Giard, medical doctor and radical parliament member Jean de Lanessan, published several influential books in which he attempted to demonstrate how the evolution of living beings brought forward a "natural" morality, which could be used as the ethical basis of the Third Republic. Another of Giard's acquaintances, Edmond Perrier (the founder of the zoological station at Saint-Vaast-la-Hougue), sought

republican inspiration in the social structure of animal colonies.[11] Such biologically inspired ideology was not foreign to the Wimereux circle either.
In the early 1900s, Pelseneer explicitly argued that acquired characteristics
would be transmitted to strengthen interclass solidarity and mutual aid.[12]

Giard himself might have hardly spoken out on the political consequences
of his science, but a single short digression hints at his stance. Interfering in
the parliamentary debates on the mine strikes in the northern-French town
of Anzin, he used biological imagery in his pleas for governmental benevo-
lence. The miners, so he argued, had been forced into an atrocious "struggle
for life" because of the performance-based wage system introduced by the
mining company. According to Giard this system boiled down to "the right
of the strong to famish the weak." This system, he suggested, was unnatural and was clearly opposed to the principle of "association for life," or "the
corrective of nature for the brutal principle of the struggle for life."[13] This
self-correcting principle could be observed among the marine creatures in
Wimereux, but could in Giard's reasoning also be a source of inspiration in
organizing the Third Republic.

If the strikes of Anzin constitute a "realm of memory" in French society,
it is not because of Giard's parliamentary intervention, but rather because
the naturalist writer Émile Zola took them as a source of inspiration for his
masterpiece *Germinal* (1885). The lives of Giard and Zola intersect, however,
and they do so in ways that are significant in understanding the position
of ecological thinking in late-nineteenth-century French society. The two
men met each other in 1883 in the Grand Hotel of Brittany's coastal town
Bénodet. Giard was there to perform comparative marine research among
others in the close-by marine station of Concarneau, whereas Zola was
finishing a novel (*Joie de vivre*) focusing on a Parisian butcher's daughter
who ended up in a degenerate French seaside village.

The nature in Bénodet, Zola happily discovered, was "wild to the point of
troubling one," and he described his isolation there as "absolute."[14] The spatial particularities were of importance, since Zola—like Giard—heavily drew
on them for his work. In Giard Zola found a like-minded spirit, who convinced him to concentrate his subsequent novel on the social conditions of
the miners of Valenciennes—the region he represented in Parliament. When
the following winter Giard explicitly invited Zola to guide him around, the
latter accepted the invitation. They ended up in Anzin, where Zola (who was
presented as being Giard's parliamentary secretary) visited the strike-ridden
mines. In many ways, Zola's documentary interest resembled Giard's. Like
the Giardist ecologist would do, Zola filled a notebook with observations

on the landscape, the food of the miners, their sexual life, and their habits. Similar themes would eventually be addressed in *Germinal*.[15]All this does not mean that Giard's ecological work directly inspired the naturalist writer. The similarities between the scientific and literary approaches of Giard and Zola can rather be attributed to an increasingly fashionable environmental determinism among the republican left. It was the same intellectual vogue that had earned Giard his Parisian chair.

THE POPULAR REALM AND APPLIED SCIENCE

Science played a central role in the worldview of the ideologues of the Third Republic, and they actively promoted science vulgarization as a way to national renewal and progress.[16] Giard, however, did not believe vulgarization offered a viable way to spread his ecological ideas. Both in teaching and publishing, he was repulsed by what he called "the modern lovers of exaggerated pedagogy, the downward-levelers."[17] As such, he would never translate his place-based research to a language suited for a broad audience. His pupils were more active in this respect. Bohn, for example, published about his scientific work in a literary magazine such as *Mercure de France* and issued books with the vulgarizing editing house of Ernest Flammarion.[18] Others specialized in holding lectures for a wide audience and leading excursions for local nature lovers. Particularly Jean Massart set up popular walking tours in several Belgian regions. In this he was obviously inspired by Giard, but, contrary to his master, he did not limit the excursion groups to a small faction of devoted pupils. In the years around 1900, Massart used the University Extension of the Brussels University as a platform to set up tours with up to one hundred participants. The longish excursion reports, published by Massart's pupils, mix light-hearted descriptions of the scenic landscape with digressions on mimetism or parasitary castration. As such they brought Giard's ecology to a large audience, helped to gain momentum for field biology, and eventually created a mainstay for the nascent Belgian nature protection movement.[19] It taught a large (middle-class) audience to look at nature with a trained eye.

To engage a lay audience for ecology, the Giardists not only guided non-scientists in the field but also translated their ideas into amateur field manuals. Cépède's project to make a French variant of *Der Strandwanderer* (the successful amateur handbook of the German botanist Paul Kuckuck) failed, but his friend Bohn did succeed in a similar venture.[20] In 1904, the latter published a reworked version of the popular nature guide by the naturalist

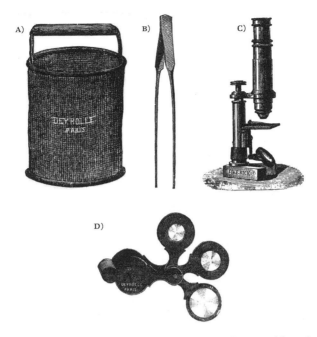

FIGURE 4.1 Field equipment to study living organisms as illustrated by Bohn: A, a cloth bucket, B, a pair of tweezers, C, a field microscope, and D, magnifying glasses. From Capus, *Guide du naturaliste* (1903), 89.

and traveler Guillaume Capus. To Capus' chapters on how to find, kill, and preserve specimens, Bohn added sections on how to observe animals in the wild and keep them alive in aquariums and terrariums. He even initiated his readership in the intricacies of "doing the tides." All these additions, Bohn claimed, were prompted by the fact that, since the book had been first published in 1879, a new science had come into being. This science, obviously, was Giard's ecology.[21]

Yet, it was not only through vulgarization, but also through practical application that ecology could prove its value to the country. Giard himself played a prominent role in developing the latter strategy. The study of the animal's relation with its milieu, he claimed, offered several possibilities for direct use in agriculture and the fishery industry. Making a geographical argument, Giard stressed that the importance of farming and industry in "his" province of Pas-de-Calais meant that he could not disengage himself from applied science. Therefore he believed it to be his duty to act as an expert in governmental bodies such as the Technical Commission for the

Study of Detrimental Insects and the Consultative Committee of Marine Fishery.[22]

In the same vein some of Giard's pupils would build out their careers in the management of natural resources. Important in this regard was the fishery station of Boulogne-sur-Mer. This station, the first of its kind in France, was set up in 1883 with the mixed support of the city council and the local Chamber of Commerce. Its tasks were to supervise the coastal fisheries, to organize the foundation of oyster parks, and to study the possibilities for repopulating the French rivers. Very close to Wimereux, Giard at first saw the institution as a competitor and he tried to get it under his personal direction. This strategy was unsuccessful, but Giard did succeed in appointing his then-assistant of the Wimereux station, Eugène Canu, as assistant manager and later full director. When Canu eventually left his directorship to enter radical republican politics, he was replaced by Adolphe Cligny, another Giardist. Both Canu and Cligny would present much of their applied fishery research as a branch of Giard's ecology.[23]

The extension of Giardist ecology to encompass agricultural and fisheries research meant that the nature under scrutiny would no longer be limited to purportedly wild places such as Wimereux. It was also to take account of the man-made farming landscapes of Pas-de-Calais and the heavily exploited fishing grounds of the North Sea. Yet, in discussing the pests that harmed French agriculture or the decline of fishing stocks, Giard (and his pupils) always used the "natural" situation in Wimereux as a point of reference. In some cases, Giard did so because he believed the parasites that were affecting northern-French crops could be traced back to similar species living on wild plants.[24] More often, however, he used his knowledge gained in the Wimereux coastal area in a less concrete and a more theoretical manner. Giard believed that unspoiled nature offered a state of "biological equilibrium" and that many of the problems facing both modern agriculture and the fishing industries were due to a violation of this natural balance. Studying how the equilibrium worked in the natural situation would give man the key to successfully manage his natural resources. The science to turn to was obviously *éthologie*, or in Giard's words: "the science of equilibriums."[25]

The causes of equilibrium disruption could be many. Giard pointed among others to abusive fishing and to the introduction of invasive species by modern transport. Yet he believed mankind could restore the balance by "imitating nature."[26] One way to do so would be to ban chemical pesticides and replace them with parasites that "biologically castrated" detrimental insects. In line with the ideas of the microbiologist Élie Metchnikoff, Giard would study the possibilities of parasitic fungi in countering scarab plagues. In Russia, Giard's acquaintance Isaak Krassilstchik set up a "small experimental factory" for the culture of fungi spores to work out the idea. In France, however, the approach was not really taken up.[27]

Giard would not limit himself to fungi spores, but formulated policy recommendations with regard to various fishery and agriculture questions. He suggested introducing intervals in the cultivation of crops so as to interfere with the reproductive cycle of harmful insects; with boulders he hoped to alter the geological milieu in which parasites of oysters thrived; the introduction of fish-ladders was to help salmon to overcome dam barriers in the French rivers; sea reserves were to ensure the recovery of overfished species.[28] All these proposed measures reflected an ecological interest in the behavior of animals, their habitat, migration, and life cycles. Giard's expertise in these matters gradually earned him a place in governmental reports and specialized farming and fishery journals both in France and abroad. Ecology, so Giard managed to communicate, was a science not just of wild places, but also of the artificial landscapes that drove the French economy.

COMPARATIVE PSYCHOLOGY AND SOCIAL ECOLOGY

Giard's applied ecology often concerned small facts regarding specific groups of animals. Yet at the same time the enterprise as a whole carried a larger promise. The implicit assumption was always that ecology would reveal something not only on the life of particular mollusks or insects, but also about the motives of living beings in general; that it not only would help to manage agricultural and fishery resources, but also would ultimately lead to an understanding of human society. Because of this promise, Giard's ecological attitude would eventually be picked up in both psychology and sociology.

The Giardists themselves described much of what they did as a form of *psychological* research. Bohn and Rabaud, for example, presented their tropism research as part of an emerging discipline to which they referred as *comparative* or *zoological psychology*. These terms were not new for that matter. The anti-evolutionist zoologist Pierre Flourens used them for the indoor studies of animal instinct he performed from the 1860s onward.[29] Flourens' program did not exactly get off to a flying start, however, and comparative zoology acquired a certain dynamism only around 1900, when it developed a more physicochemical approach and incorporated evolutionary theory. Although the aging Giard took part in this movement, the center of it would be not his station in Wimereux, but rather the newly founded General Psychological Institute in Paris.

The Institute was established in 1900, on the initiative of an attaché of the Russian embassy, Serge Youriévitch, who convinced various leading figures in French psychology, such as Pierre Janet and Charles Richet, to join the venture. In origin, the institution was designed to study the psychological phenomena witnessed at spiritualist séances, but within a few months it broadened its perspectives to bring in all aspects of psychology. Rather quickly, state officials and administrators became more important in the organization, which led to a growing interest for the role psychology could play in governmental issues. It also infused the Institute with the "scientistic" and "solidarist" conceptions which were en vogue in the Third Republic—starting from the hope that psychology could provide scientific principles to structure society on a cooperative basis. This early intellectual reorientation went together with a compartmentalization in various independent research units in 1902. One of the most vigorous of these units was the research group devoted to zoological psychology. It was in this group that the Giardists would play an important role. Apart from Giard, Bohn and Rabaud set the tone at its gatherings.[30]

In the early days of the institute *zoological psychology* and *comparative psychology* were still rather ill-defined terms. Giard, Bohn, and Rabaud seem to have used them particularly for studies of the movement of animals, their senses, preference behavior, and habits of orientation. Gatherings of the research group discussed issues such as the role of different sensations in the behavior of hermit crabs, the subconscious of self-raised birds, or the origins of the maternal instinct.[31] The first full-length memoir published by the group was a work of Bohn, which, on the basis of field experiences in Wimereux, studied the influence of light on the movement of marine animals.[32] In all these cases the issue of the interaction between the organism and the environment was raised, so that the exact boundaries with *éthologie* remained unclear. In general, Giard and his pupils seem to have used the term *zoological psychology* mostly for the (physiological) study of automatic reactions and instincts, while *éthologie* referred to broader issues of adaptation and the interrelation between various species.

The General Psychological Institute was founded to throw light on the psyche of humans, and the study of animals was included only for the comparative help it could provide in interpreting human psychology. In this way it offered the Giardists the possibility to tout the wider relevance of their

FIGURE 4.2 A postcard issued by the General Psychological Institute, entitled "Science un-
veiling nature." The topos was an old one (dating back to at least the seventeenth century),
but it certainly fit well with the ideology of the Third Republic in which science was propagated as a new religion. (Bourgeron–Rue des Archives / The Granger Collection)

work. The clearest example of this is probably Giard's study on the origins
of maternal love. In this he tried to understand the phenomenon in humans
through extensive comparisons with various animal species. Systematically,
Giard pointed to the physiological origins of maternal behavior—albeit these
could widely differ depending on the organism's relation to its milieu. In
humans, he argued, the relation between child and mother was very comparable to that of the parasite and its host. As in many parasitic relations
there was a mutual equilibrium offering both partners an advantage. This
equilibrium was not unchangeable. Motherly love, he indicated, had altered
through evolution. He hoped it might evolve further, away from primitive
egoism and mere familial altruism, into a phase of universal unselfishness.
In this phase, mothers would raise future generations as a community. As
a model, Giard was thinking of social insects such as bees, a species that

saw sterile females raising the offspring of others. The whole argument brought several key elements of Giardist thought together. It wedded studies of Wimereux invertebrate parasites to the behavior of women in the Third Republic, comparative ecology to evolutionary theory, and (somewhat ambiguously) natural determinism to voluntaristic politics.[33]

After the death of Giard in 1909, Bohn was the Giardist who went most public as a comparative psychologist. At the same time he never let go of his ecological interest. What distinguished French comparative psychology from other similar traditions, Bohn argued, was precisely its ecological method. This method had been developed for biology, but could be easily incorporated in psychology as well. Its greatest advantage was that it opened the eyes to the whole complex of environmental factors that interfered in animal behavior. In Bohn's view, taking these factors into account would make it possible to rule out "hypothetical faculties" such as the organism's "will." Bohn himself particularly explored "vital rhythms," such as responses to rhythmic variations in the milieu such as tides, and "differential sensibilities," such as reactions to sudden environmental changes. It was such subjects that constituted the core of his "ecological" comparative psychology.[34] At the same time, like Giard, he would stress that the importance of his research went beyond this limited number of invertebrates. The deterministic explanations of their behavior, he insisted, could "be extended to higher animals, and even Man."[35]

All in all, the success of Bohn's comparative psychology was limited in scope. The practices connected to it never really seem to have traveled outside of his own laboratory at the École des Hautes Études in Paris. In the United States—the country that according to Bohn himself was most active in the study of animal psychology—his works were regarded with suspicion.[36] In France itself, the tropism concept met a lot of resistance and Bohn was not able to maintain his good relations with compatriots interested in animal psychology. Especially after the First World War, he quickly became marginalized.[37] In the meantime, however, his works had been read in broad intellectual circles and occasionally his ideas were picked up by scholars from other disciplines. The most notable example of this transfer of ideas can be found in the work of Émile Waxweiler—an engineer who was to become the leading figure in Belgian sociology between the turn of the century and the year of his untimely death in 1916.

Waxweiler was not the first sociologist to be interested in the work performed in Giardist ecology. As early as the 1890s, Jean Massart had used his own ecological experiences as a point of departure to think about society.

Like many biologically inspired thinkers of the time, he showed a strong organicist bent, seeing (or at least metaphorically representing) society as an organism.[38] Together with the sociologist and later socialist party leader Émile Vandervelde, Massart published an article in 1893 in Giard's *Bulletin* under the title "Parasitisme organique et parasitisme social" ["Organic and Social Parasitism"]. In this paper Massart and Vandervelde compared the functioning of parasites as observed in Wimereux with the life of "social parasites" in human society. The latter could be persons of private means, thieves, or prostitutes. The article met with critical acclaim and the authors were eager to extend the success.[39] In 1897, they joined ranks with the physiologist Jean Demoor to publish *L'évolution regressive en biologie et sociologie* [*Regressive Evolution in Biology and Sociology*]. This book was again based on an organicist comparison: the evolutionary regression of organs and the historical waning of social institutions. As in the case of organic and social parasitism, parallels and differences were considered in great detail.[40] In both works, one might argue that Massart and his coauthors used ecology not so much as a scientific model, but rather as a supplier of metaphors. Waxweiler, then, wanted to go a step further. His inspiration in Giard's ecology did not concern metaphorical language: it concerned a method. This method, so he argued, not only differed from that of organicist sociology, but actually undermined it.

The center from which Waxweiler led his attack on organicist sociology was the Institute of Sociology in Brussels, founded in 1902. In close contact with his patron, the wealthy industrialist Ernest Solvay, Waxweiler gave the institute its theoretical and material basis. He personally drew the plans for the new building, and he attracted collaborators from varying disciplinary backgrounds. Rather quickly, he came up with an elaborate theoretical framework to unite his multidisciplinary staff in a common cause. In 1906, he published *Esquisse d'une sociologie* [A sketch of sociology], a book that would serve as the bible for the work done in the institute for the following ten years. In its central postulate the echo of Giard's work is clear. It stated that sociology could be nothing but "*éthologie sociale*," or "social ecology."[41]

Waxweiler defined *social ecology* as "the physiology of the reactional phenomena due to the mutual excitations of individuals of the same species without distinction of sex." The definition contains the central idea of Waxweiler's critique of organicist thinking. The only tangible object of study for a sociologist was concrete individuals, he argued. The social organism was merely a creation of the mind, a construct not susceptible to objective analysis. In his search for an alternative, Waxweiler took Bohn's and Loeb's tropism research as a model. Just like them, he wanted to focus on individual reflex behavior (or in his own terms: "reactional phenomena"). Although the definition left open the possibility of studying all kinds of organisms, Waxweiler himself was particularly interested in humans.[42]

With his comments on the prevailing sociology of his time, Waxweiler followed Giard's criticism of microscopic biology. He argued that much of sociology was merely "descriptive"—based on statistics of "dead facts" or even "fossils of facts." His ecological method had to bring life to the discipline, by refocusing on the present-day reactions of living individuals. Unlike the organicist sociologists, he did not want to reconstruct the historical evolution of social organisms; what he aimed at was the analysis of the actual mechanism of social change. In this, he took over Giard's neo-Lamarckian bent. Individual adaptations, Waxweiler argued, could become fixed in "hereditary habits." This mechanism, he believed, could be observed both on the biological and the social level.[43] All in all, however, Waxweiler and most of his collaborators did not stress the weight of the hereditary past, but rather the adaptiveness of humankind. In their view, it was not hard heredity but the changeable milieu that accounted for human behavior. Hence, by changing the milieu one could change men—whether these were the lower classes in Western societies or the indigenous people in the colonies.[44] Just like Giard's ecology, Waxweiler's social ecology was an optimistic science.

In Belgian sociology Waxweiler's expanding institute occupied a central position and within its walls the concepts of its director were ubiquitous. Not surprisingly, the references to ecological terminology abound in the work of his collaborators. Yet underneath the terminological layer old traditions survived. The human scientists of the institute introduced terms like *éthologie* and *adaptation to the milieu*, but hardly changed anything substantial in their existing research traditions. Some of the life scientists on the other hand showed difficulties in combining Waxweiler's framework with older more deterministic lines of thought.[45] It is not much of a surprise, then, that the program of social ecology collapsed almost immediately after Waxweiler's death in a car accident in 1916. In the postwar context it was the more practical concerns of the reconstruction that would become the focus of the Institute of Sociology. For this task, sociologists no longer seemed to need Giard's conceptions.[46] The journey of his ecological attitude in Belgian sociology abruptly came to an end.

MUSEUM PRACTICE AND ZOO DESIGN

Of more lasting influence than Waxweiler's sociological program was the incorporation of Giard's ecological ideas in the field of paleontology. The man to adapt these ideas to the interpretation of age-old fossils was the French engineer and naturalist Louis Dollo, another former student of Giard. After studies in Lille, Dollo was engaged in the Brussels Royal Museum for Natural History in 1882. It was here that, in the following thirty years, he would develop his particular conception of *"paléontologie éthologique"* or "ecological paleontology."[47] In this way, the Giardist approach would cross another disciplinary border.

From the day he started at the Museum, Dollo published an almost end-less series of short analytical papers. The more programmatic texts in his oeuvre were few, but influential. They included "La paléontologie éthol-ogique" [Ecological paleontology] (1909), the only article in which Dollo re-ally elaborated on the goals and methodology of his discipline. Paleontology, he stressed there, was not to be considered as a part of geology (as was usually the case), but rather as a subdiscipline of biology. After all, for Dollo, the major aim of the paleontologist was the reconstruction of evolutionary trees. This orientation, of course, was not entirely new. Scientists such as the London zoologist Thomas Huxley had clearly influenced this line of thought. What Dollo added, however, was the ecological element. For recon-structing the tree of life, morphology was not enough, Dollo argued. After all, common characteristics of fossils could be explained not only by com-mon descent, but also by convergent evolution due to adaptation to similar environments. To distinguish between the two, one had to bring the pre-sumed ecology of the studied organisms into account. In this way he trans-lated the approach of his former master to the study of fossils.

Giard followed Dollo throughout his career, gave him advice, supported some of his better-known theories, and offered him space in his *Bulletin*.[48] At the same time, Giard's expertise in paleontology was limited and Dollo remained a bit of an outsider in the Wimereux network.[49] Giard's ecology could be incorporated into the world of the paleontologists, but not with-out changing its character. Unlike Giard, Dollo was after all not a man of the field. He was a cabinet scientist who has been described as "a virtual recluse with almost monastic habits."[50] In ecological paleontology, the study of the animal's milieu was not to be done in nature itself, but rather in the mind. This is already clear in the very first assignment Dollo fulfilled at the Museum: the reconstruction of the recently discovered iguanodons of

Bernissart. Up to the 1880s, these dinosaurs had always been presented as ponderous quadrupeds.[51] Dollo however put them in an upright position, arguing that it would have given these swamp animals an evolutionary advantage to spot predators over great distances. Standing on their hind legs, they could seize possible attackers with their short but powerful forelegs and hurt them with the horns on their thumbs. Unlike paleontologists before him, Dollo thus studied the possible advantages of morphological characteristics in a concrete milieu.[52]

It was logic rather than field observation that was Dollo's forte. It was also logic that was behind his most famous contribution to paleontology: the law of the irreversibility of evolution. This law was very much connected to his ecological interest. The question concerned in Dollo's law was whether an organism that returned to an ancestral milieu would turn completely back to its ancestral morphology. If this was the case, it would be almost impossible to reconstruct evolutionary trees—as the direction of evolution could always be turned around. According to Dollo, however, evolution was not reversible. To return to a form that was completely identical to the ancestral one, the organism had to undergo all the ecological changes that had influenced its constitution, but in reverse. By all laws of probability this was impossible, according to Dollo.[53]

Dollo's work concerned the interpretation of bones of long-extinct animals. His approach, however, coincided with a more general Giardist worldview. In courses he gave for a local boarding school, he mixed sessions on the iguanodon with more general lectures. In these he indicated that life was "the result of constant exchanges between the living being and the environment." This insight concerned not only morphology, but also psychology. Human behavior, after all, was to be interpreted as a series of reflexive responses to the environment, without the intervention of a free will. Changes of the milieu, he insisted, left their traces in contemporary marriage, crime, and suicide rates. The environment, Dollo told his young students, shaped not only the body form of iguanodons, but also the behavior of every single one of them.[54] Ecological paleontology, it is clear, was part of a larger philosophy of environmental determinism.

Originally, Dollo's ecological paleontology received little acclaim in the French speaking world, where paleontology would largely stay an affair of the geologists.[55] Outside Belgium and France, however, his research program met with a growing popularity in the interwar years. The Austrian paleontologist Othenio Abel would eventually set up the journal *Palaeobiologica* after an event organized in honor of Dollo's seventieth birthday. Dollo died a few months after the first issue was published, but thanks to the journal his scientific program survived up to the Second World War. In 1948 the journal stopped, as a result of financial problems, but from the 1970s onward its ideas were revived in the work of "radical paleontologists" such as Stephen Jay Gould. It is certainly no coincidence that the latter proved to be a great fan of Dollo.[56]

Whereas Dollo introduced Giard's ecology in the museum, it was another of Giard's pupils, Gustave Loisel, who tried to integrate its concepts in the zoological garden. Loisel had frequented Giard's laboratory in the Sorbonne, and although he was not a visitor at Wimereux he picked up the ecological sensibilities of Giard. At the beginning of the twentieth century, he would distinguish himself with a series of proposals to reform the infrastructure of the French zoos from a Giardist point of view.[57]

In 1902, Loisel worked out a first ambitious plan to set up a biological farm for "experimental transformism" at Giard's laboratory in Paris. This proposal came to nothing. Then, in 1905, partly to create a professional position for himself, Loisel developed a plan to reorganize the menagerie of the Muséum d'Histoire Naturelle. This did not lead to direct results either, but Loisel was nonetheless noticed by the Minister of Instruction. A year later, the latter entrusted him with a series of missions to study the design of zoos in Europe and the United States. The long reports which resulted from this study tour were again followed by a worked-out plan to create a zoological garden of a new kind. The plan—for which Loisel collaborated with the architect Désiré Bessin and the engineer-zoologist Albert Chappelier—was exhibited at the Salon of Architecture in 1908.[58] Although this project (again) came to nothing, it is revealing for the ways in which it merged Giardist ecology with a new role of the zoo in managing the natural resources of the country.

Loisel designed an institution that combined scientific research with popularization, but without being "sensationalist." Furthermore he wanted the institution to focus on (indigenous) animals in their natural milieu. Just as Bohn had stressed that local field knowledge was necessary for devising a successful laboratory set-up, Loisel believed such knowledge was indispensable for conceiving a proper zoo. He took some inspiration from the well-known zoological garden of Carl Hagenbeck in the German town of Stellingen, but he disliked its stress on showmanship.[59] Instead of putting energy into painted snow peaks and trompe l'oeils, one should rather invest in scientific infrastructure, Loisel argued. Therefore, the artificial hill he designed as the center piece of his zoo offered "ideal" conditions not only for animals and visitors, but also for scientists. The hill consisted of several parks, in which animals of different species lived in semi-wild conditions. Both on its top and at its bottom, there would be observatories, where scientists could observe the animal's natural behavior and, centrally, there would be a well-equipped laboratory. Yet the stress of the research of Loisel's zoo would be on ecology, variation studies, and biogeography. In all this, it was observation rather than manipulation that was to be the main task. If the zoo staged experiments, it should be "those which are used by nature itself," not "the violent and sometimes even cruel means which are used in the laboratories of physiology and experimental medicine." The echo of Giard's thinking is clear.[60]

The importance of the study of "nature itself" in the zoo partially lay in its possible translation into matters of "economic zoology." The latter was the crux for the managerial role Loisel had in mind for zoological gardens. He believed the study of animals in the (semi)natural environment of the zoo would empower scientists to interfere successfully with the more complex nature of the nation as a whole. The zoo would offer a place to study the effect of nourishment on the productivity of animals; it would be a site where one could research how to efficiently domesticate and acclimatize animals; and it would be the ideal setting to investigate how to conserve useful wild species and destroy species that were considered harmful. It would be a laboratory for improving the use of the nation's animal resources.[61]

Loisel's ambitious zoological projects did not trigger any concrete result. He became embittered and after World War I he left the world of biology completely to devote himself to the historical study of the Roman emperor Marcus Aurelius.[62] His plans had definitely been among the most imaginative to emerge from Giard's entourage, but he never raised enough support to get them realized. Unlike the often very cheap, ecological practices that his colleagues had developed for the field, Loisel's plans to set up similar research in the more controlled surroundings of the zoo required an expensive infrastructure. Loisel did not have the necessary connections to provide him with funding for such a project, nor could he rely on the resources of a well-established discipline.

Wimereux was without doubt the center from which Giard's ecological attitude spread. It was there that the bonds of Giard's clan were forged, where his ecological *topoi* took shape and its methods were worked out. When Giard entered his Parisian laboratory, the French Parliament, or a cornfield

A)

B)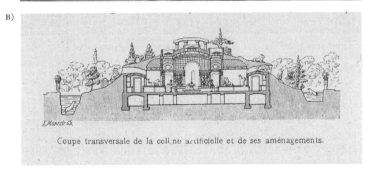

Coupe transversale de la colline artificielle et de ses aménagements.

C)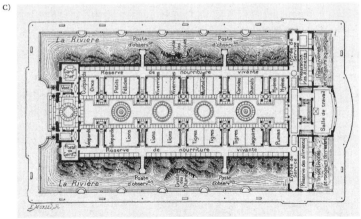

FIGURE 4.3 A–C, Gustave Loisel's design of a scientific zoo. From Hachet-Souplet, "Un projet de transformation" (1909), 179.

in Pas-de-Calais he kept referring back to concepts that had been developed in his zoological station. The same is partially true for his pupils, who brought Giardist ecology among others to the university of Lille and the Sorbonne, the General Psychological Institute, the Institute of Sociology and the Natural History Museum in Brussels. The group of people who engaged in this was a diverse one. They were not united by the boundaries or the infrastructure of a clearly delimited discipline, nor by an institutional linkage or a shared type of workplace. They shared a way of looking, rather than a particular subject. *Éthologie* was an "undisciplined science."

Not all projects that sprung from Giard's circle were successful. And yet, at the beginning of the twentieth century the Giardists had conquered an important place in the intellectual landscape of French-speaking Europe. This was partially due to the success of Giard's clan itself. According to the literary critic Marcel Coulon, the Giardists made up two-thirds of "the official milieu" in 1910 and even three-quarters in 1925.[63] Whatever the value and precise meaning of this statement, it is true that many of Giard's followers had successful scientific careers. To be sure, the ambitions of some Giardists were frustrated (e.g., in the case of Loisel), but others rose to prominent chairs, curatorships, and directorships. Because Giard's pupils integrated his ecological viewpoint in their widely divergent activities, they were able to leave a mark on how the French and the Belgians thought about nature as such.

What Giard tried to do in Wimereux was understand how organisms interact with their environment. In this chapter, I indicated how this project mapped onto the wider cultural aspirations of turn-of-the-century France and Belgium. The Giardist clan belonged to a group of late-nineteenth-century bourgeois who were convinced determinists, evolutionists, and free-thinkers, and who sided with the politically progressive. The study of small invertebrates in Wimereux tide pools received meaning in that social context. This study was perceived as a first step in revealing the laws of life (now and in the distant past), understanding contemporary society, and managing the natural world. Studies on organisms in the Wimereux area trained the eye for interpreting other organisms in other contexts. The success of Giard's ecology in the French-speaking world can be ascribed to weak disciplinary boundaries, strong personal ties, and (partially unarticulated) ideological connections. For all these reasons Wimereux could become a reference point for a myriad of scientific ideas and practices far beyond the beaches of Pas-de-Calais.

CHAPTER FIVE

Plön

A Lake Microcosm

The rise of the biological field station prompted a new way of looking at nature. This way of looking was characterized not only by its novel methods and ideas, but also by its geographical focus. The earliest biological stations brought the immediate surroundings of the coastline into the center of the zoologist's attention. Visitors to stations such as the ones in Naples and Wimereux concentrated their work on a small geographical zone that did not exceed a few kilometers beyond the shoreline. In fact, before 1890, zoological stations were used to explore nothing but this particular environmental space.

Only in the margin of the station movement did some zoologists delve into habitats that could be encountered in other geographical realms. One of these habitats that originally triggered no more than a marginal interest was the inland lake. Seemingly a logical extension of the study of marine fauna, freshwater zoology constituted only an infrequent side project for the early station zoologist. Giard's clan, to use this example again, almost entirely neglected the subject. Just Jules de Guerne, one of Giard's early pupils, took up a real interest in the topic, but without finding support from the government or the academic elite. But by the end of the century, things would change. In 1892, De Guerne reported in the *Revue scientifique* that a permanent station devoted to freshwater research had been founded in the northern-German town of Plön. According to De Guerne, the station, established by the plankton specialist Otto Zacharias, opened up "a promising path" in the history of freshwater zoology. He hoped other nations (including his own) would soon follow the German example.[1]

The station of Plön was set up as part of a larger research agenda that addressed the biological study of lakes (or: limnology). Today, this interest constitutes a successful branch of ecology, and limnologists are generally acknowledged to have contributed important insights to contemporary ecological theory.[2] It was the station of Plön that, particularly from the 1920s onward, served as the European center of this innovative and increasingly acclaimed research. Also elsewhere, the rise of limnology was closely connected to that of its prototypical workplace: the lakeside field station. For these reasons, the history of freshwater stations and the science performed there might easily be told teleologically. One could with De Guerne present it as the account of a "promising path" that lay open and that simply had to be followed. In fact, such a path never existed.

The early history of freshwater zoology reveals that, rather than an easy walk on a sign-posted path, it was a struggle to make inland lakes and their animal inhabitants into objects of scientific study. New practices and rhetorical strategies had to be developed, a new scientific tradition had to be pieced together, and a new class of scientists (and consumers of their science) had to be constructed. Also freshwater zoology itself was an object of construction. Zacharias might have described his workplace as a *zoological* station, but the work performed there was not framed within clear disciplinary boundaries. Zacharias and his followers saw the biological study of lakes as a project that held a mediating position between various sciences such as chemistry, hydrography, botany, and zoology. The story of the early days of the Plön station reveals the intricacies of such a transdisciplinary project. It involved a lot of frustration, ad-libbing, and changes of direction. The whole enterprise was one of struggle indeed—even a "struggle for life," if one would want to use Zacharias's own somewhat dramatic phraseology.[3]

Zacharias not only wanted to launch a particular type of plankton research in Plön. Like Giard, he expected to spread his approach and hoped it could also appeal outside of its original context. Several factors offered promises in this regard—among which an increasing German interest for the nature of the *Heimat* (or: homeland), the rise of applied fishery research, and a burgeoning school reform movement. This chapter will explore how these factors stimulated and shaped Zacharias's work. I will argue that they accounted for part of the success of freshwater zoology in the civic milieu. At the same time we will see that Zacharias failed, much to his annoyance, to enthuse academia for the work of his station.

A JOURNALIST IN HOLSTEIN SWITZERLAND

When the freshwater station in Plön was inaugurated in 1891, Zacharias was already forty-five years old. For him, directing the station meant the start of a second career. Zacharias's first calling had been that of science popularizer and journalist, and it is particularly for his work in that capacity that he is still remembered today. This background is of significance. As we will see, Zacharias's personal history is indicative for the early years of freshwater zoology as a whole. The study of freshwater animals was, after all, a science that largely originated in the world of amateurs and practical naturalists, rather than that of academic heavyweights. This origin was of importance for the functioning of the Plön station. It left its marks in Zacharias's personal networks and his reputation, his scientific ideals, his frustrations and self-image. For all these reasons, it is useful to have a short look at Zacharias's pre-Plön career.

The first time Zacharias was really noted by the German scientific elite was in the mid-1870s, when he vainly attempted to put out the first German monthly for the propagation of the theory of evolution. The journal Zacharias had in mind would offer a "popular yet comprehensive" discussion of all things Darwinian. The elderly Darwin himself approved of the plans, and Zacharias quickly made headway. By the spring of 1876 the young journalist had gathered the consent of a large group of prospective collaborators, including the leading German Darwinian of the time, Ernst Haeckel.[4] A title—*Darwiniana*—was agreed upon, contacts with publishers were arranged, and a first issue was prepared. Yet, in the course of the events, Haeckel (who was to serve as the figurehead of the journal) had growing doubts about Zacharias's scientific and editorial abilities. With the excuse of fatigue the Jena zoologist kept Zacharias off, while setting up a project for a periodical of his own. The latter would bear the title *Kosmos* and have roughly the same collaborators as Zacharias's intended monthly. The only major difference with *Darwiniana* was that *Kosmos* would have Haeckel's acquaintance Ernst Krause and not Zacharias as its editor in chief. Once Zacharias found out about Haeckel's maneuvers, it came to a breach between the two. The first issue of *Kosmos* appeared in March 1877; *Darwiniana* was never to be published.[5]

The "putsch" against Zacharias revolved around his scientific authority. To one of his close friends, Haeckel explained that Krause was to be preferred because he was "more skilled."[6] Krause himself, then, was even harsher in his opinion of Zacharias: "the man is as uneducated as he is unsuited." With "uneducated" Krause referred to the fact that Zacharias had not received an academic training in the natural sciences. Krause might have been a science popularizer himself, but his training as a pharmacist seemed to have enhanced his standing. Zacharias for his part was what Haeckel described (in his more supportive letters) as a "self-made man."[7] After having attended the Sunday school of the Polytechnic Society in Leipzig, Zacharias had started his career as the mechanic of the university's astronomic observatory. Only later he took up an academic study of metaphysics—a subject he purportedly tried to master in cafés, with French and English lexicons on his lap. After his graduation in 1869, Zacharias tried to make himself a career as a journalist. The plans of putting up his own Darwinian journal were only an extension of this journalistic activity. Zacharias's credentials in evolutionary biology were still rather limited at the time. His knowledge of zoology could largely be traced back to the reading of popular Darwinian books and some hands-on (but autodidactic) experience with the study of microscopic fauna. Apparently this did not suffice to convince Haeckel and the like that he was suited to lead the first German journal devoted to evolutionary theory.[8]

The *Darwiniana* debacle did not temper Zacharias's scientific ambitions. In its aftermath he continued to profusely write popularizing articles. At the same time, he showed an increasing ambition to prove himself as a scientist in his own right. In 1883 he started to attend public lectures of the leading zoologist (and anti-Haeckelian) Rudolf Leuckart, and he participated in anatomy practicals at the university of Leipzig. The year after, he acquired funding from the Prussian Academy of Sciences to carry out freshwater zoological research. Zacharias's letterheads of 1885 self-importantly mention a "private laboratory for microscopic and zootomic research" in the Silesian town of Hirschberg as his workplace. This laboratory—most likely not more than a room in his house—served as a hub for exploring the lakes in the nearby mountain range. Despite limited support, Zacharias managed to publish the results of his excursions in elite science journals such as *Zeitschrift für wissenschaftliche Zoologie, Biologisches Centralblatt*, and *Zoologischer Anzeiger.*[9] The journalist reinvented himself as a "proper" zoologist. In a letter in which he (unsuccessfully) attempted to restore his relations with Haeckel, he wrote: "I have completely burned all the ships that could lead me back to the bleak coast of journalist life, and competition with the type of people that dwell at such coasts no longer exists."[10]

The foundation of a permanent station for freshwater zoology was only the next step in Zacharias's self-created career as a scientist. Setting up

A)

B)

C)

FIGURE 5.1 Zacharias re-enacting scientific activities in a photo studio. A, teaching, B, reading, C, microscoping. (Courtesy of the Max Planck Institute for Evolutionary Biology, Plön)

this station was a matter of prestige and credibility. The marine zoologists had shown in their sparring with the physiologists to what extent material workplaces could be helpful in acquiring a reputation of scientific relevance and progressiveness. Setting up a station for freshwater zoology, Zacharias hoped, would similarly increase the standing of his particular field of enquiry. He also hoped it would help to restore his own reputation.

Setting up a station was not a simple task. It involved a great deal of networking, negotiating, and fundraising, and the preparations took Zacharias at least four years. All this time, Anton Dohrn's privately owned Stazione

Zoologica in Naples served him as a model for what he wanted to achieve.[11] Eventually, he also managed to mobilize the same networks that had been instrumental in the foundation of the Stazione. Friedrich Alfred Krupp, the arms manufacturing giant who financed several of Dohrn's projects, donated a large sum for Zacharias's station. And like Dohrn, Zacharias maintained good contacts at the Optical Workshop of Carl Zeiss, which would freely supply him with scientific instruments. Like the project in Naples, the station in Plön furthermore relied on the goodwill of the local policymakers. Zacharias was able to convince the mayor of Plön to supply him with a station building—as, indeed, Dohrn had been granted a building site by the Naples City Council. Most important, however, was the sponsorship by the Prussian Ministries of Culture and Agriculture. To favorably influence these, Zacharias could rely on the same politically influential scientists who had been supportive of the Stazione: Rudolf Virchow in the Prussian Parliament and Emil du Bois-Reymond in the Academy of Sciences in Berlin.[12] The fact that all these men shared a personal animosity toward the controversial Haeckel only enhanced the bonds. Berlin, Naples, and Plön soon were regarded as mutually supportive strongholds in an anti-Haeckelian front. The first letter of du Bois-Reymond to Zacharias after the founding of the Plön station is illustrative in this regard. It opens: "Let us not worry about Mister Haeckel and his acolytes. He is a professional liar and a slanderer."[13]

Despite the continuity in the networks behind the stations in Naples and Plön, it is important to see that the latter originated in a different context than the former. The plans of the Stazione had been set out in the period before the Franco-Prussian war and had carried a rather cosmopolitan undertone. By the time the freshwater station in Plön was founded two decades later, Germany had been unified in one Empire and there was a growing consensus on the *national* importance of science—as a source of both prestige and economic applications. In this context, Zacharias was eager to stress that the zoological knowledge of "native freshwater lakes" was at least as important as that of "some southern European gulf." Such local knowledge, he stressed, would in any case be of more value to the German fishery industry.[14] With this type of rhetoric he particularly addressed the German fishery societies, which from the 1880s onward showed an increasing preparedness to sponsor zoological research.[15] And it worked. The West-Prussian Fishery Union opened the pages of its journal for Zacharias's fundraising articles and donated to the fund that was to defray the cost of the station's equipment. The list of other donors gives a further impression of the strong local foundations of Zacharias's enterprise. It includes several regional natural history societies, next to the individual names of local noblemen, industrialists, and—an echo of Zacharias's first career—publishers.[16]

More even than the possibilities offered by the local civic milieu, Zacharias was attracted by the ecological characteristics of the Plön area. The town was located in a region that not only was particularly rich in freshwater reservoirs, but also had them in a great variety of forms, ranging from large lakes to small ponds and fen pools. Zacharias himself stressed that one could reach nineteen big lakes and a diverse range of smaller water reservoirs in a car ride of only three hours. This variety implied that the freshwater zoologist could compare the animal world of very differing habitats without covering too much distance, and, as such, the Plön region seemed especially suited as a "natural laboratory." Lakes of very different morphologies could be selected and compared, for instance, with regard to their plankton richness. The diverse landscape, as such, almost naturally suggested particular practices of place. Overall, Zacharias's reasons to settle in Plön were quite similar to those that, twenty years earlier, had made Alfred Giard decide to set up his station in Wimereux.[17]

In the patchwork of Holstein waters, it was one lake in particular that eventually drew Zacharias: the Großer Plöner See. This lake not only was large, but also had an irregular bed that showed important differences in depth, thus offering another source of habitat variety. The fact that the lake had no big inflows and, hence, was relatively isolated, was seen as an additional asset. As we will see, Zacharias would often argue that the clear ecological boundaries of lakes allowed the freshwater zoologist to gain general biological insights unreachable for biologists specialized in less surveyable habitats.[18] Zacharias's Großer Plöner See, thus, could serve as a model for the clearly bordered habitats of the lake as such. However, it epitomized not only the "universal lake," but also a particular regional variant of it. In his publications Zacharias claimed that the Großer Plöner See could serve as a "prototype" of basically all the large lakes one found in the northern-German lands between Mecklenburg and the Russian border. These lakes clearly differed from the much better studied Alpine ones, which were deeper and poorer in animal life. By constructing a different, "northern" type of lake with the Großer Plöner See as its model, Zacharias could carve out his own scientific niche and play the regional card at the same time. The fact that the Großer Plöner See counted as "unpolluted" was obviously crucial as well. In this way the "northern" freshwater fauna could be studied in its supposed natural state.[19]

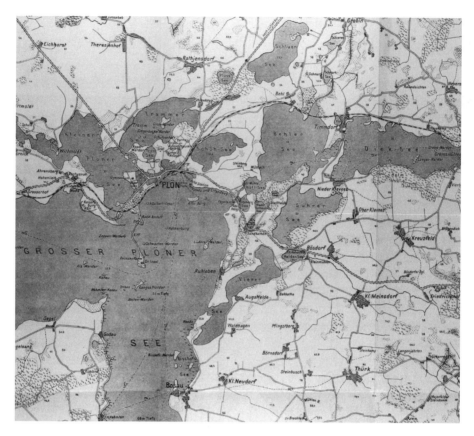

FIGURE 5.2 The Plön region: nineteen lakes, fens and ponds within a three-hour car drive. From Zacharias, *Das Plankton* (1907), 217.

In choosing a site for his station, Zacharias desired not only an unspoiled habitat that suited various species of freshwater animals, but also an environment that pleased the visiting zoologist. Inspired by Dohrn's example in Naples, Zacharias was convinced that researchers would want to combine their work with pleasure and rest. Holstein, which increasingly counted as a bourgeois holiday destination, was a good choice in this respect. In the latter decades of the nineteenth century, the lake area around Plön witnessed the development of a series of typically German (somewhat understated) lakeside resorts, and the region became known as a place of healthy hikes and steamship lake-outings. Zacharias certainly saw the potentiality of this type of middle-class romanticism. In his articles, he started to describe the region as "Holstein Switzerland" (a denomination that was a travel guide invention of the 1850s), and he explicitly mentioned the cheap-yet-comfortable hotels, the beautiful walking trails, and the closeness of a lakeside resort. He also eagerly quoted the *Berliner Tageblatt* that strongly approved of his geographical choice: "A king could not choose a more splendiferous spot on this earth, when, tired of the hustle and bustle, he would desire to spend some happy days enjoying a terrific landscape."[20]

Late-nineteenth-century Holstein was not only a place of happy stillness and postcard landscapes, however. It was also a region in the midst of rapid changes. A long-time part of Denmark, the (largely German-speaking) territory passed to Austria in 1864 after the Second Schleswig War to be ceded to Prussia two years later. After its integration in the Reich in 1871, the region would continue to play an important symbolic role in the German national mind as one of the recently acquired borderlands. Scientifically conquering the lakes of this area could obviously be seen as an extension of its territorial annexation, just like the founding of the biological station in Helgoland (an island taken over from the British in 1890) served as a marker that the isle was an integral part of the Reich. By making the Holstein lakes prototypical for the whole of northern Germany, Zacharias moreover made its political annexation into a "natural" fact.

The political changes also left material traces in the Holstein landscape. The take-over by the Prussians went together with a somewhat belated industrial revolution, triggered among others by an unprecedented railroad boom. At an increasing pace a growing network of railways was constructed, connecting the province to the major cities in the south. Zacharias's project largely depended on this new infrastructure. Since 1866, Plön had a railroad connection with Kiel; since 1873 with Lübeck. These connections brought its picturesque lakes within the reach of urban biologists with little time. Hamburg, so Zacharias advertised, was only two hours away from his station; Berlin eight, and Leipzig ten. At least in travel hours the metropolis had come relatively close to Holstein Switzerland.[21]

Modernity showed its face in other ways as well. Much of Holstein remained rural, but also the countryside lost many of its long-established customs. Intensive root crops, steam tractors, and artificial fertilizers were introduced; the surviving heath, marshes, and moors disappeared; local handicraft was competed out of the market by urban industry. In the meantime, the relative importance of agriculture and the fisheries—Holstein's key economic sectors—declined. Some country people profited from the growing regional markets, but many others were forced to migrate to the cities or overseas. In Kiel, Holstein's major city, the changes were even more intense than on the Holstein countryside. Originally known as a small and idyllic *Landstadt*, its population quadrupled in the three decades after the Prussian annexation. In 1865 Kiel became the Prussian navy headquarters, which besides marines brought a steadily growing shipbuilding industry. At the same time, the city also slowly gained a national reputation as an intellectual center. Its university developed a new dynamic, increasingly hiring professors from other parts of the Reich rather than local savants. The economic changes in Holstein as a whole disturbed the social fabric. A sense of loss was felt among large parts of the population, explaining a growing nostalgic interest in the *Heimat*, its folklore, landscape, and nature.[22]

Zacharias stood amid these changes and was determined to make use of them. The expanding Kiel university, so he suggested, could provide his station with resources such as scientific equipment, literature, and assistants, and he hoped his freshwater research would profit from the Prussian aspirations to modernize the local fisheries. Yet, while embracing modernity, he would also explicitly present freshwater zoology as an aspect of the increasingly popular *Heimatkunde* (homeland studies), offering the prospect of reconnection to an eternal and regionally embedded nature. In his attempts to put a new field of research on the agenda, Zacharias thus easily mixed nostalgia with modernism.[23]

FRAMING THE LAKE

When Zacharias opened his station, freshwater zoology was still very much of a marginal science, hardly taken up at the German universities. In 1889, Zacharias himself described it as "the Cinderella of zoological research." The expression would reappear as a mantra through the rest of his work. Freshwater fauna, so he lamented, was a subject thought to be worthy only of "the Wednesday and Sunday excursions of aquarium lovers." Those who engaged in it, he added, were "not considered to be equal members of the research guild."[24] The aquarium lovers to whom Zacharias referred were hobbyists, who had largely taken on the subject in the wake of the 1857 publication of *Das Süßwasser-aquarium* [The freshwater aquarium], a popular book published by the journalist Adolf Rossmässler. Yet, while introducing the freshwater world in the domestic sphere, Rossmässler's work failed to have a substantial influence on elite science.[25] In contrast to the fauna of the sea, which since midcentury was increasingly studied by professional zoologists, the world of lakes and ponds continued to be considered, again according to Zacharias, as nothing but "poor, monotonous and dull."[26]

Also internationally, freshwater zoology was still in its infancy when Zacharias took it up. It was not until the 1860s that some coordinated interest in the topic developed. In those years, a handful of Scandinavian researchers started to counter the idea that lakes were zoological "wastelands" by uncovering a hitherto neglected world of microscopic crustaceans commonly known as water fleas.[27] It was in the Switzerland of the 1870s, then, that the field would receive its major impetus. This was largely to the credit of Alphonse Forel, professor at the Academy of Lausanne. Forel's interest concerned the "oceanography of lakes": the study of the physical and chemical characteristics of freshwater reservoirs and the interaction of these characteristics with the lakes' fauna and flora. To describe this field of study, Forel coined the term "limnology." Next to this coinage, his early work was of importance for its indication that freshwater organisms existed at a much greater depth than generally assumed and for its delineation of several "lake regions" with differing life conditions. His eventual magnum opus, the three-volume *Le Lac Léman* (1891–1904), would remain a work of reference for many decades.[28]

Zacharias had a high regard for Forel, but developed his project largely independently from Lausanne.[29] He also hardly used Forel's term "Limnologie" to refer to his own activities. In fact, there was a certain indecisiveness in Zacharias's terminology with regard to his own discipline. It was clear that his personal interest was in the microscopic fauna of lakes, but he was dithering in his strategies to frame this interest in a larger discipline. Often he used the term "Hydrobiologie," which referred to the study of both aquatic animals and plants that live in salt water as well as freshwater. To indicate his special interests for lakes, however, he frequently added the adjective "lakustrisch" (or: "lacustrine"). Alternatively, he also described his discipline as "Planktonkunde" (or: plankton studies). *Plankton* was a term coined in 1887 by the Kiel physiologist Victor Hensen, who used it to describe the "floating plant and animal material in the sea."[30] As the word quickly included the small free-floating organisms in lakes, "Planktonkunde" could be seen as a subdiscipline of "Hydrobiologie." When referring to his own station, Zacharias used neither term, however, mostly just describing it as a "lacustrine zoological station."

The choice of a discipline's name was a weighty matter. Changing terminology meant including or excluding the study of certain habitats and organisms. When macroscopic organisms were included, this could lead to an association with the economically important study of fish; when sea habitats were included, an alliance could be set up with the academically very successful work of marine zoologists. It is clear, however, that above all,

Zacharias wanted freshwater plankton research to come out of the shadow of more successful bordering research programs. He might have sometimes strategically associated his projects with marine zoology or applied fishery research, but his main goal was to organize a discipline around the study of freshwater microscopic organisms. Much of his work, in fact, reads as an apology of such a discipline.

Repeatedly Zacharias would formulate and reformulate arguments to convince German academia that lake plankton was a worthy research object. He argued among others that planktonic organisms were particularly interesting from an evolutionary perspective, since these primitive creatures could be helpful in uncovering the origin of life.[31] Furthermore, when studied under the microscope many of these proved to be of an intense beauty. This, Zacharias would stress, was of scientific importance. After all, he believed that "the intensive sense of well-being, triggered by Nature's beauty, is a factor that affects Man as a whole and empowers him with physical and intellectual capacities, which, under normal circumstances, are not at his disposal."[32] This idea, which gives the aesthetic sensation of nature a central role in its intellectual understanding, was obviously not new. It can be traced back to the romantic conceptions of Humboldt and Goethe and it was well alive in the late-nineteenth-century work of Haeckel and Wilhelm Bölsche.[33] The novel thing Zacharias offered was a new source of aesthetic delight: the planktonic creatures of the lake.

Zacharias's rhetoric was influenced by that of the marine zoologists, who had been so successful in conquering a place in late-nineteenth-century biology. Yet, for limnology this rhetoric would be only partially effective. After all, in many aspects the sea seemed to outshine the lake. Its fauna was richer and more diverse and it was more capable of triggering the imagination of evolutionists than that of inland waters. Zacharias was eager to stress the hypothesis that the primeval sea might have been free of salt, so that the source of life should be sought in lakes rather than the sea, but he had to admit that phylogenists found many more "intermediary forms" among marine species than they did in freshwater fauna.[34] To make matters worse, sea creatures seemed to generate a much stronger aesthetic sensation in turn-of-the-century Germany than freshwater plankton did. Compared with the exotic and opulent organisms shown in Haeckel's *Kunstformen der Natur* [*Art Forms of Nature*], Zacharias's microphotographs showing freshwater algae indeed looked a bit meager.[35]

Hence, if Zacharias wanted limnology to come out of the shadow of

Fig. 2. Leptodora hyalina (Planktonkrebs).

FIGURE 5.3 *Leptodora hyalina* or "plankton crab." From Zacharias, "Das Plankton" (1897), 699.

marine zoology, he had to look for arguments other than the aesthetics, diversity, or evolutionary importance of freshwater animals. He primarily searched these arguments on the practical level. Oceans might impress the human sentiment, he wrote in 1895, but, because of their extension and the heterogeneity of their fauna, they were hard to survey as a whole. The lake to the contrary, with its clear borders and its limited number of species, offered a much more surveyable situation for those who wanted to unravel the household of nature.[36] Was the lake not "a world on its own" that provided a look at nature as such? Its spatial particularities, Zacharias suggested, made it a research site that was preferable to the sea.

From the early 1900s onward, Zacharias would use the term "microcosm" to describe the well-delineated freshwater world of the lake.[37] Zacharias was obviously not the first to call this metaphor for the defense of limnology. The American Stephen Forbes had already famously done so in his 1887 paper "The Lake as a Microcosm." Yet Zacharias gave the metaphor a particular meaning. Unlike Forbes, he connected it with an optimistic

Humboldtian nature image, which was still very popular among German civic zoologists. Rather than Darwinian struggle, so central in Forbes's interpretation of freshwater life, Zacharias stressed harmony and balance.[38] He even made the study of lake harmonies into the central focus of his discipline.[39] This holistic approach had a clear spiritual side to it. In his earliest articles on freshwater research, Zacharias had already stressed how the study of lakes would make humanity conscious of "its own connection with nature."[40] As such, the lake microcosm was seen as pedagogically interesting. In Zacharias's view, excursions to local lakes were the ideal way for students to connect to nature and learn the basic mechanisms of its household. This was also practically possible, since, unlike seas, one could find lakes almost everywhere close to hand. Next to objects of study they offered places of spiritual connection to the *Heimat*.[41]

With such rhetoric, Zacharias was not so much echoing academic developments, but rather continuing an old discourse of popular biology. In the midcentury Rossmässler had already shown the same sensibilities, when urging his bourgeois readers to fill their aquariums with regional rather than exotic species.[42] In the 1880s, then, a similar *Heimat*-approach had been successfully defended by the Holstein school director Friedrich Junge in his widely read reform book *Der Dorfteich als Lebensgemeinschaft* [The village pond as a living community]. It argued that by studying the interconnected lives of the organisms in a local village pond, schoolchildren would learn much more than by just memorizing taxonomic categories.[43] Zacharias followed suit. Not long after *Der Dorfteich* was brought out, he published one of his first articles on freshwater collecting in a book with the telling title *Anleitung zur deutschen Landes- und Volksforschung* [Introduction into German land and people research].[44] In later work he indicated that studying local lakes was of great importance since it would "indirectly increase the love for the fatherland and the interest for the direct environment."[45]

In Zacharias's eyes, freshwater zoology thus earned a place next to the existing biological disciplines, because of its conceptual potential, its holistic appeal, its pedagogical promise, and its presumed links with *Heimatkunde*. Furthermore, he hoped the study of lakes could serve as a lever to force a general reorientation in the zoological research of his time. In terms very similar to those used by the French Giardists, Zacharias pleaded for a return to free nature. And, as Giard had done, Zacharias decried the one-sided laboratory orientation he thought to be typical of his time. "Laboratory researchers," according to him, were often mesmerized by "the smallest details," in this way "losing their sense for the animal world as a whole and for its aesthetic appreciation, reaching a one-sidedness, which one can no longer justifiably describe as a science of the animal kingdom, but at most as [a science] of the fine structure of the animal body."[46] Next to that narrow laboratory science, he believed there was ample room for other approaches.

Zacharias's strategy bore close resemblance to that of other proponents of a "new natural history." Their attacks did not focus on laboratory biology as such, but on special branches of it, such as histology. This choice of target was highly tactical. Much of the work performed in the latter discipline was descriptive, so that it was easy for Zacharias to portray field-based limnology as the more "experimental" of the two. In this way the usual associations connected to the lab and the field were cleverly turned around. The reversal was further strengthened metaphorically by representing free nature, in Giard's line, as one big laboratory.[47] In Zacharias's case the metaphorical language was particularly worked out to serve the case of limnology. "Every lake," he wrote in 1888, "is a research aquarium for breeding experiments performed by nature itself."[48] The lake was controllable just like a laboratory device, but at the same time—and unlike laboratory devices—it offered a look at the *Gesammt-Natur*. Free nature was not just any laboratory, it was "the most encompassing" one.[49] And it was by far preferable to the closed quarters of the histology lab.

Next to the histologists, also the evolutionary morphologists were the object of Zacharias's criticism. This can be seen as a way to get even with Haeckel, the best-known representative of the discipline. In any case, it was an attempt to reclaim evolutionary theory for field biology. Evolutionists, Zacharias and his collaborators would stress, should not limit themselves to reconstructing evolutionary trees on the basis of sliced-and-stained organisms, but they should also study the actual adaptation of animals to their environment. Had this after all not been Darwin's first interest?[50] And wasn't the field, rather than the morphologist's cabinet, the place where one had to pursue it? German zoologists, Zacharias believed, had become numb because of too much indoor work. It was this point he eagerly hinted at in his last letter to Haeckel. The short note, written in 1898, concerns a zoologist who wanted to come to Plön as an assistant, carrying Haeckel's recommendations. Zacharias had no money to hire him, but he nonetheless asked Haeckel whether the man would be suited for the job. Somewhat stingingly, he added: "I do not conceal from you that I have already received [...] people with a complete and first-rate histological training, who, with regard to freshwater fauna (and flora), acted like helpless children."[51]

As in Giard's defense of ecology, much in Zacharias's argumentation in favor of limnology concerned space. In his representation, limnology appeared as a prototypical outdoor science. It was no science of laboratory recluses nor of old-fashioned "museum zoologists." Both the laboratory and the museum detached specimens from their natural habitat, whereas, according to Zacharias, one had to study them in their own "dwelling place." Only there, he believed, one could come to understand their living, feeding, and reproductive habits, the fluctuation of their numbers, their life cycles, their instincts, and the impact of the environment on their variation.[52] Zacharias stressed this work was important—despite a widespread contempt for "those who tire themselves out to establish the ecology of all the organisms, which the slice-series-makers melt in their paraffin blocks or fix on their microscopic slides."[53]

Zacharias's allegiance to fieldwork might surprise. His own expertise, after all, concerned plankton, which is obviously too small to be observed "in the wild." Much of its study necessitated microscopes or weighing apparatuses and actually happened inside the station. Zacharias's heroic accounts of ecologists "enduring rain and sunburn" also lose some of their power when one knows that his own fieldwork was partially outsourced to local fishermen.[54] To be fair, Zacharias's field rhetoric was never without ambiguity to begin with. Although he was eager to associate his project with fatiguing and adventurous excursions, he also defended his station for the comfort and control it offered to the visiting scientist. Field excursions were important, he believed, but they did not suffice to build out a solid science. This could only be done from a permanent station.

Before the establishment of his station, Zacharias had carried out many improvised field trips himself. Hence, he knew the bother that came with furnishing hotel rooms as laboratories and the difficulties brought by the transport of boats and equipment. He noted that, since most excursions were short, one usually spent them feverishly collecting, without having time to study the behavior of animals in their natural habitat. Once back home, then, the researcher always realized he had brought back too many specimens of certain groups and too few of others. Zacharias furthermore stressed that field excursions tended to be carried out during summer holidays, so that little, if anything, was known about freshwater life in winter. Overall, he stressed that knowledge based on excursions alone had a coincidental character, which could only be remedied by daily, year-round observations carried out from a fixed station. Zacharias admitted that for comparative reasons work performed in and around the station should be supplemented with excursions to other regions, but he insisted that these would be fruitful only if coordinated from one single center. More even than Giard, Zacharias stressed the importance of longevity. If one wanted to understand the complex processes of nature, one had to observe the same phenomenon over long stretches of time. What was lost in geographical breadth was won in temporal scope.[55] The scientist, Zacharias believed, should not approach nature as an agitated traveler. Rather he should make it his home.

THE KIEL CONNECTION

From the 1880s until his death four decades later, Zacharias endlessly reiterated that the field-oriented ecological approach remained neglected among German university zoologists. At the same time, he saw exceptions to the general rule. The universities of Berlin, Leipzig, and Kiel, so he indicated, developed into important (albeit isolated) centers of the "biological perspective" that he pursued himself.[56] Zacharias hoped that his own limnological project could eventually infiltrate academia via these centers. This seemed a reasonable aspiration. Zacharias's scientific program, after all, had clear links with those of the zoology departments in the respective universities. He drew on theories and methodologies developed there, albeit he modified these to suit a freshwater context. Particularly the influence of Kiel, next to which Zacharias's station was located, was very strong. It was above all there that Zacharias would look for academic input for freshwater zoology.

The university of Kiel was Germany's foremost center of ecological research. Its ecological dynamic dates back to 1868, when the university established its first chair of zoology. The man hired for the job, Karl August Möbius, was not an academic heavyweight and mostly had a background in popular biology. He had worked as a teacher in a primary and later a secondary school; he had been engaged in the founding of the Hamburg zoo and its aquarium; he was known for his science journalism and popular lectures. Yet, at the same time, Möbius had built out his own research expertise. Together with his friend and supporter Heinrich Adolf Meyer, he had set up a large-scale study of the fauna of the Kiel Fjörd. The resulting book, *Die Fauna der Kieler Bucht* [*The Fauna of the Kiel Bay*] (1865 and 1872), was novel for its sophisticated treatment of "societies" of organisms, which it related to the physical conditions of the geographic zones in the Fjörd. The book clearly showed Möbius's ambition to be taken seriously not only as a popularizer but also as a researcher in his own right. In this sense, the Möbius of the 1860s strongly resembled the Zacharias of the 1880s.[57]

Möbius's appointment in Kiel in 1868 not only gave him the recognition he yearned for, but also offered him the position to parade the economic importance and the theoretical bearings of his ecological approach. In response to a request of the Prussian Ministry of Agriculture, he started a wide-ranging research on the depletion of oyster banks in Schleswig-Holstein. The results were published in 1877 as *Die Auster und die Austernwirtschaft* [*The Oyster and Oyster-culture*]. In this book, Möbius introduced the term *Lebensgemeinschaft*, together with a more academic expression for the same concept: *Biocönose*. In Möbius's definition a *Biocönose* was "a community of living beings" or "the mutually dependent species and individuals that maintain themselves in a definite territory by means of reproduction, and, the selection and number of which is determined by the average external life conditions."[58] A change in any factor of the *Biocönose* entailed changes in all the other according to Möbius. This explained why the overfishing of oysters in France had led to their partial replacement in the life community by cockles and mussels. The *Biocönose* concept, thus, seemed to offer a tool to handle the lingering crisis in the oyster industry, but its echo was also wider. In the world of popular biology, the *Lebensgemeinschaft* would quickly become a household term, thanks (among others) to Junge's aforementioned *Der Dorfteich*. Among university zoologists the rise of the concept was somewhat slower, but its eventual success was nonetheless conspicuous. Today *Die Auster und die Austernwirtschaft* counts as one of the foundational texts of the discipline of community ecology.[59]

Möbius was not the only initiator of the ecological approach that was developed in late-nineteenth-century Kiel. His younger colleague, Victor Hensen, played an equally important part. Hensen was in many respects Möbius's antipode. From a well-to-do family background, Hensen had taken medical studies in several leading universities and he had quickly climbed the academic ladder after graduating. In 1864, at the age of twenty-nine, he was appointed as extraordinary professor of physiology in Kiel. The difference in background between Möbius and Hensen left its traces in their respective research styles. Hensen's biological ideals did not reflect the holistic interests of popular biology, but rather the reductionist theories that were in vogue among the leading German academic physiologists of the time. His approach to ecology, then, would be mostly quantitative, and in this way complement the largely qualitative concepts of Möbius.[60]

For both Möbius and Hensen, state support was crucial for developing their research projects. In 1870 both entered the Commission for the Investigation of the German Seas, newly set up in Kiel by the Ministry of Agriculture. Its assignment was to collect data on the physical and chemical characteristics of the sea and the life habits of the organisms that lived in it. In this context, a series of scientific observatories was set up along the German coast and three subsequent oceanographic cruises were organized. Hensen stood at the center of these enterprises. After having focused his own attention on fish and fish eggs, he increasingly concentrated on what he described as the metabolism of the sea. Central in understanding this metabolism, so he believed, was plankton, the group of organisms he named. By quantitatively studying the spread of plankton he hoped one would be able to unravel the cycle of matter in the seas. Hensen therefore put a lot of effort into devising nets and laboratory equipment to quantify the mass of plankton in a given volume of water. With this new equipment and a worked-out methodology he left Germany in 1889, accompanied by five other Kiel scientists, to perform the three-month Plankton-Expedition that would cross the Atlantic Ocean. In the same year, Zacharias published his first articles that announced the founding of his freshwater station at only 30 kilometers from Hensen's institute.

Zacharias's own ideas about understanding the fauna of lakes closely matched those that the leading zoologists in Kiel had developed for the sea. Yet, Zacharias seems to have acquired them somewhat independently. Like Hensen, he was used to thinking about organisms quantitatively. In Zacharias's case, however, it had not been reductionist physiology, but rather Darwinian theory and Malthusian thinking that had been inspirational. And it was not so much the study of plankton that had put Zacharias on the path of quantification, but rather his analysis of human society. Crucial in the latter regard was a pamphlet he published in 1880. The pamphlet was devoted to the famine that swept Upper Silesia, Zacharias's place of residence at the time.

Late-nineteenth-century Upper Silesia was a region known not for picturesque nature, but for coal mining, overpopulation, child labor, and hunger typhus. A report by Rudolf Virchow on the typhus epidemic in 1848 had brought the region's poverty to the center of political attention.[61] The situation of the province continued to be problematic in the following decades, and when in 1879 a food crisis struck Silesia again, this was generally seen as a disgrace for modern Prussia.[62] Yet Zacharias believed that in the ongoing debates the voice of science remained unheard. In his pamphlet of 1880 he argued that neither state-enforced wage increases nor an extension of the school system would do much to counter the problem. Referring to Malthus and Darwin, he claimed that the high birthrate was the ultimate cause of hunger in the region. The Silesian struggle for life, so he believed, could only

be tempered if the average family size went down.[63] Such demographic reasoning might seem very dissimilar from the Hensen-style plankton research
that Zacharias would perform later, but in fact the two had a lot in common.
Both focused on populations (rather than individuals), on interdependence,
and on food chains. Furthermore *Planktonkunde* was intimately connected
to the problems Zacharias signaled in his 1880 pamphlet. Plankton after all
became a source of state interest because it fed fish, and because the fish
fed humans. The quantitative study of plankton was to a large extent about
feeding the rapidly growing population of modern Prussia.

The fact that Zacharias's zoology was rooted in somber reflections about
Silesia's demography does not mean he stuck to a conception of nature
that was red in tooth and claw. We have already indicated that, besides a
Malthusian inspiration, his conception of nature was also influenced by
Humboldtian ideas of natural harmony. It was this aspect that eventually
got the upper hand when Zacharias moved away from the overpopulated
Silesian mining regions to the idyllic lakes of Holstein. The shift prepared
him to appreciate the work of Möbius, who was another great admirer of
Humboldt and whose theories of life communities combined easily with
a harmonic conception of nature's economy. In the late 1880s, Möbius's
Biocönose concept had already found its way to Zacharias's articles.[64] The
latter, thus, seemed well disposed to integrate both the quantitative and the
qualitative aspects of the Kiel program into the research agenda of his future station. Such a station in nature, so it seemed, offered even better prospects for this program than the urban-based laboratories in Kiel itself.

Zacharias took the first steps for a concrete collaboration with the Kiel
school in 1886, when he introduced himself to Möbius. The latter seemed
to get along well with the ambitious zoologist, and the two men kept in
contact.[65] There had been no talk of a freshwater station yet, when, in 1887,
Möbius left his professorship in Kiel for the directorship of the Natural
History Museum in Berlin. In the capital, Möbius noticed an increasing interest for freshwater zoology, and in February 1888 he wrote to his Kiel
successor Karl Brandt that money would be made available for research on
a still to be chosen inland lake. Möbius sidestepped Zacharias and asked
Brandt whether there were young researchers in Kiel who were suited
for the job. Brandt suggested one of Hensen's PhD students: Carl Apstein.
Almost simultaneously, Zacharias contacted Brandt about the latter's possible interest for his (still embryonic) station plans. Brandt saw possibilities
for a partnership and answered that the project interested him very much
indeed. Academic and civic limnology seemed to be heading for a close collaboration. Brandt assured Zacharias that he and Hensen would do "all in

their power" to support the project. Zacharias, for his part, suggested that his—still to be founded—station would be an ideal place for Kiel students to work on their dissertations.[66]

The contacts between Kiel and Plön led to quick results. As soon as 1891, Zacharias could proudly publish a handbook on freshwater zoology for which Apstein had written a chapter. The book could be seen as the sign of a growing coalition between the civic milieu and the university. Its editor might have been a former journalist, but most of its collaborating authors were academics. The book's projected audience was both science students and serious amateurs.[67] This readership was introduced to the most novel sampling techniques in Apstein's highly detailed chapter, whereas Zacharias initiated them into some general aspects of the relations between the lake and its fauna.[68] The same division of labor between the young Kiel Doctor and the elder civic zoologist was maintained once the Plön station was actually established. In his journal Zacharias reported that the station's "quantitative" sampling would be taken up by Apstein, while he would take care of the "qualitative research."[69] The latter term signified a whole variety of things, ranging from the inventorying of lake organisms to the study of their variations and their means of spreading from one lake to the next. Together, Apstein and Zacharias seemed to cover the whole field of an emerging discipline.

The alliance did not last long, however. Apstein's presence in Plön quickly led to a territorial conflict. In the summer of 1893, after only a few months of collaboration, Apstein published a scathing article in the local natural history journal *Die Heimat*, in which he openly attacked the station and Zacharias's work. Among others, he stressed that the station drew hardly any visitors and that Zacharias lacked the proper background to professionally lead it. In Apstein's view Zacharias could "only in a limited sense be regarded as a representative of zoology, [. . .] [but] certainly not as one of its leaders, not even in the area of freshwater research." Zacharias was said to lack love of truth and respect for the intellectual property of others. The fact that Zacharias had used techniques developed by Hensen and Apstein without properly referencing them, was taken as an insult.[70] Apstein did not fail to send offprints of the article to leading personalities in Plön and to several German zoological institutes. Zacharias, understandably, was furious. He wrote an incensed letter to Brandt and he banned Apstein from further access to the station. Brandt answered more or less diplomatically, but clearly sided with Apstein.[71] The cooperation between Plön and Kiel stopped short. Civic and academic limnology, again, largely went their own ways.

FIGURE 5.4 The Plön station: designed as a meeting point between civic and academic science. (Courtesy of the Max Planck Institute for Evolutionary Biology, Plön)

PURE SCIENCE AND *BILDUNG*

After the clash with Haeckel almost two decades earlier, the *Heimat* incident offered a second occasion on which Zacharias's ability was put into question from an academic side. Also this time, he reacted combatively. Among others, he published a pamphlet in which he gathered letters of leading zoologists, such as Rudolf Leuckart and Carl Chun, who spoke out positively about his station.[72] This public support did not, however, guarantee a large influx of academic zoologists. Unlike many marine stations, Plön would not become the center around which an academic research clan gathered. Zacharias listed only twenty-one guests in the first five years, and among them only three professors.[73]

This does not mean that the Plön station was unable to develop its own research dynamic. The house journal *Forschungsberichte aus der Biologische Station zu Plön*, the first periodical devoted to freshwater biology and ecology in the German-speaking world, published yearly heavy volumes. In order to fill these, Zacharias relied more on practical naturalists than academic scientists. Ernst Lemmermann, for example, the journal's algae specialist, was a teacher and, later in his career, scientific collaborator at the Bremen Museum für Natur-, Völker-, und Handelskunde; Emil Walter was the director of the fishery station in Trachenberg, but also editor of a fishery journal and author of popular fishery books; Max Voigt had taken zoology courses at the university of Leipzig, but eventually became a teacher and a popular science writer; Hugo Reichelt, finally, was a tradesman, diatom specialist, and prominent member of the Leipzig natural history society.[74] Together these naturalists represented a broad spectrum of zoological expertise, despite the fact that their profiles differed drastically from that of the average author who published in journals such as *Zoologischer Anzeiger* or *Zeitschrift für wissenschaftliche Zoologie*. Men such as Lemmermann, Walter, Voigt, Reichelt, and indeed Zacharias himself created a wide-ranging visibility for limnology. They did so through popular lectures, museum exhibits, and articles in general journals and periodicals for applied science. Yet, they clearly had more difficulty in reaching the German academic community.

All this does not mean that the research performed in Plön was very different from that promoted in Kiel. Despite his breach with Brandt, Zacharias himself largely continued the type of studies he had outlined in 1891. Among others, he took over the sampling project of Apstein, looking into the seasonal mass variations of plankton, and analyzing its horizontal and vertical distribution.[75] These methods eventually put him in a position to detect dissimilarities between differently shaped lakes with respect to nutrient richness and plankton mass.[76] As Giard had done in Wimereux, Zacharias developed a particular comparative method to study the interaction of animals with their environment in Plön. Unlike Giard, however, he compared not individuals, but entire lake microcosms. It was a practice of place that was made possible only through Kiel-style quantification. At the same time, however, Zacharias stressed the relativity of Hensen's and Apstein's counting methods. Unlike the plankton researchers in Kiel, he believed that plankton formed "swarms" rather than being distributed uniformly in large volumes of water. This implied that sampling results were much more coincidental than Hensen and his followers were ready to admit. The measurement of "the total production" of a lake, Zacharias indicated, would therefore always remain a bit of "a lottery."[77] Probably because of this conviction, his main interest remained qualitative. Next to some faunistic research, his work would particularly focus on the periodic appearance of freshwater animals, their seasonal dimorphism, and their biogeography.[78]

Because academic recognition failed to appear, one might have expected Zacharias to reorient to applied fishery research—for which the civic milieu generated most funding.[79] Yet he continued to profile his station as a place of "pure" science, stressing that "general and far-reaching research" could not be directed by fishery interests.[80] He indicated that applied fishery research had its reason of existence, but that, ideally, it was separated both intellectually and geographically from limnology. Such a division of labor, Zacharias insisted, actually was taking form in Germany, when, in 1894, the aforementioned fishery station in Trachenberg was founded. This new station built out its program in dialogue with Plön. Although Zacharias continued to do some research into the feeding habits of fish, he would pass the practical elaboration of his findings to his Trachenberg colleague. Pure science was thus granted its separate sphere, while at the same time it was represented as an indirect source of economic applications.[81] Such a rhetoric of "indirect use" was obviously widespread among proponents of pure science. For Zacharias it was a strategy to maintain both the intellectual independence of his project and its appeal to politicians and fishery societies.

Despite his rhetoric of pure science, Zacharias's station largely failed to acquire the aura of scientific professionalism its director strove for so vigorously. A change of journal title from *Forschungsberichte* to the more inclusive *Archiv für Hydrobiologie und Planktonkunde* did not change the tide either, and Zacharias increasingly showed his bitterness. The station had been a long-time "problem child," he claimed in 1906, "looked upon distrustfully by many and seen as a bastard between science and dilettantism."[82] The frustration was clear, and ultimately led Zacharias to reorient his station's focus once more. From a place of limnological research he would, from around 1905 onward, attempt to turn it into a center of pedagogical reform.

Just as Zacharias's scientific project was rooted in a discontent about the one-sided biology performed in (most) universities, so his educational mission originated in unhappiness about ongoing pedagogic developments. Traditionally, German nineteenth-century education had been closely associated with the concept of *Bildung*. Although the actual meaning of the term evolved through time, its use in the educational discourse mostly involved a stress on self-cultivation, personality building, and thematic broadness. By the time Zacharias was setting up his project of freshwater zoology, many commentators believed these values to be in decline. An increasing specialization and utilitarianism in the German education system was thought to undermine its traditional *Bildung* values. Cultural critics of various kinds used this gloomy image to launch their own proposals for revival. Some connected a restoration of the neohumanist *Bildung* ideal with an anti-positivist attack on science.[83] Others, like indeed Zacharias, believed that it was exactly science that could reinvigorate the *Bildung* of Germany's youth. It was not dead languages, Zacharias stressed, but living nature that provided "tangible and clear objects" for the students to be "absorbed in the soul." According to Zacharias, plankton in particular would make the ideal "object of an up-to-date biological school education."[84]

Zacharias's plankton-centered idea of school reform might seem the project of an isolated eccentric, but in fact it belonged to a well-established and increasingly influential movement. After Junge had published his *Dorfteich* in 1886, a growing group of German freshwater zoologists published reform books. *Das Leben der Binnengewässer* [Life of the inland waters], issued by the museum curator and school reformer Kurt Lampert in 1899, was particularly influential and paved the way for a whole series of popular publications dealing with microscopic freshwater organisms. Zacharias published his popularizing *Das Süsswasser-Plankton* [Freshwater plankton] in 1907; his acquaintances Arthur Seligo (director of a fishery society) and Raoul Francé (leader of the Royal Hungarian Society for Natural Sciences) came with comparable books. The virtually untranslatable title of the latter's booklet nicely sums up the central credence of the group: *Der Bildungswert der Kleinwelt*.[85] The Wilhelmine citizen, so the freshwater zoologists believed, would come to self-cultivation and synthetic knowledge by peering into the microcosm of a nearby lake. As such, they were in tune with much of the cultural criticism en vogue among German turn-of-the-century intellectuals. Their sensibilities are clearly expressed in Zacharias's hope that student lake excursions would counter Germany's prevailing materialism and the suicides generally associated with it.[86]

Zacharias believed that his role in the announced pedagogic revolution should be more than issuing popular books and writing contemplative articles. He actually wanted to initiate the teachers of the German *Gymnasiums* and *Realschule* into the basics of plankton research. For this purpose, Zacharias replaced the occasional seminars he had been giving with a full-fledged summer school in hydrobiology and plankton studies. He convinced the Prussian Ministry of Culture to fund a pavilion to be built next to his Plön station, and, in 1909, he could announce his first three-week summer course. The same spatial assets that earlier had to attract professional zoologists to Plön now had to lure teachers. With learning and research rooted in a similar kind of direct field experience, it was not so strange that the same landscape could act as the ideal place of both activities.[87]

The summer school was only a partial success. As early as the second year of operation, unhappy participants voiced their critique in several educational journals. Later, a derogatory article in the leading *Zoologischer*

FIGURE 5.5 The station as a place of school reform: Zacharias and teachers posing in a purpose-built teaching pavilion. (Courtesy of the Max Planck Institute for Evolutionary Biology, Plön)

Anzeiger was published, written by zoology student Adolf Rieper. The same critique returned in all articles: the station in Plön lacked the appropriate equipment and assistants, and the focus of the courses was too narrow. Zacharias was said to have initiated his students only in the morphological study of plankton, hardly touching upon the topic of *Biocönose* that had been so prominent in the rhetoric of his new pedagogy. Participants had been highly displeased, and a purported third of them left before the end of the course.[88] The critique (again) enraged Zacharias. He responded with a sharp article in *Zoologischer Anzeiger* and an even more scornful piece in his own periodical. Admitting that his station was not luxurious, he stressed that the summer courses were meant for teachers, not "gentlemen, who come from well-funded state-institutes with high demands." Surprisingly, Zacharias admitted his morphological focus, but he tried to keep up his image of a reformer, stressing that he preferred hands-on experience with a well-chosen topic over an encyclopedic overview of the whole field. Aggressively, he made fun of the megalomania of the "Royal Highnesses Seminar Teachers," and stressed that the Plön station had already been founded when Rieper was "still in his nappies."[89]

The dispute left Zacharias particularly bitter. After having been found wanting by a leading zoology Professor (Haeckel) and a young Doctor (Apstein), he had now been forced to an unpalatable polemic with a twenty-three-year-old student. The summer school was carried on the following years, appreciative teachers published positive reviews, but it did not really live up to Zacharias's ambitions. Plön failed to become the nationwide center of school reform, just as it had failed to be the center of a new academic discipline.

Yet the early story of the station of Plön was not only one of frustration and failure. Despite the bitterness of some polemics, Zacharias had managed to use his station to substantially heighten the visibility and the standing of limnology in the German-speaking world. Together with befriended authors, such as Lampert, Seligo, and France (who all explicitly referenced Zacharias and his station), he gave the study of freshwater organisms an important boost.[90] After Junge's work had introduced it in the teachers' circuit in the 1880s, Zacharias managed to considerably broaden the group of practicing naturalists that took up the subject. Around 1900, freshwater zoology was perhaps still a Cinderella at universities, but it was no longer the exclusive domain of amateurs or individual schoolteachers either. Research into limnology and *Planktonkunde* would be associated with museums, fishery societies, and natural history associations, and in many of these contexts the station of Plön counted as a center of the profession. Furthermore, outside the German borders, Zacharias's work had inspired the foundation of several limnological stations. In foreign elite journals, the station in Plön was referred to as "the most important" and "the best known fresh-water station in the world."[91] And, although the initial ambition to turn Plön into an instrument to reform academic zoology might have remained unfulfilled during Zacharias's lifetime, this would be realized not long after his death.

INTO ACADEMIA: THIENEMANN'S PLÖN

Zacharias died on 2 October 1916, only one day after the freshwater station of Plön had celebrated its twenty-five-year existence. The last years of his directorship had not been very remarkable. The dream of establishing a discipline that focused on the ecological study of lakes was largely gone. Zacharias still published, but mostly on matters of cytology and histology—the laboratory disciplines he had criticized so heavily earlier in his career. His cytological research did not concern local species, but a universally used laboratory animal: the horse parasite *Ascaris megalocephala*.[92] It seemed the ultimate denial of his credo that biology needed to return to the local perspective, to the study of organisms in their place-specific natural habitat.

At the moment of Zacharias's death, the station of Plön was not in a particularly good state. Its equipment was out of date and its library was in disorder. Frogs and grass snakes inhabited the cellar.[93] Despite its apparent state of decline, however, the Prussian Ministries of Culture and Agriculture and several fishery societies believed the station should be maintained. In 1917, an arrangement was made so that it could be taken over by the newly founded Kaiser Wilhelm Society for the Advancement of Science, which grouped a series of state-sponsored research institutes led by elite academics exempted from teaching.[94] The Münster limnologist August Thienemann was hired as the new director, and he would lead the station for the following thirty-three years. His background was completely different from Zacharias's. Thienemann was far from an autodidact, having studied biology with leading professors at the universities of Greifswald, Innsbruck, and Heidelberg. Not long after having defended his PhD, he entered the Agricultural Research Station of Münster, where he set up a Department for Fishery and Drainage Research. Only a few years later, he became a lecturer at the university of the same town. Not at all an outsider like Zacharias, Thienemann knew the ins and outs of the academic world. The two men had occasionally met, but they never were particularly close. When Thienemann took up the directorship in 1917, he made clear that he wanted to put the station on a new footing. And he did so very effectively.[95]

Unlike Zacharias, Thienemann could successfully act as a prominent member of the academic community. Illustratively, when negotiating about the directorship of the Plön station, one of his demands was that his pay would equal that of a full professor. It was furthermore part of the deal that the directorship of the station would be combined with an extraordinary professorship at the university of Kiel. Thienemann restored the contacts with Karl Brandt and was eventually even invited to succeed him, but he chose to continue his work in Plön. In 1918 the station received its first permanent assistant and in 1928 a second, while several Kiel zoology students worked there to prepare their PhDs. The number of visitors who left a note in the station's guestbook quickly rose. Significantly, the International Society for Theoretical and Applied Limnology, founded in 1922, would have its seat in Plön. Despite the worrying state of the German economy, Thienemann was able to free increasing resources for his limnological work and the functioning of his station.[96]

Thienemann's work earned him an international reputation. This standing was based, among others, on his typology of lakes and on the introduction of the concept "production" in aquatic ecology.[97] Although researchers like Apstein had already performed limnological research in a university context, it was Thienemann who really brought the discipline to German academia. During his directorship, Plön became a leading limnological research center, housing a growing group of academic biologists and serving as an international reference point for the profession. Historians (albeit mostly German ones) have called Thienemann the founder of ecological limnology or even scientific ecology as such.[98] In historical narratives, Zacharias has therefore largely disappeared in the shadow thrown backwards by his successor. Thienemann himself played a part in this, stressing in the obituary of Zacharias that the latter's work had mostly been "descriptive." Ten years later he would stress that Zacharias's importance was to be sought "not so much in his research as in the propaganda he made for his discipline."[99] All this was a belittling kind of praise, hardly hiding a feeling of superiority.

To be fair, Thienemann published far more than Zacharias and, in general, his work was theoretically more ambitious and chemically better informed. Yet despite Thienemann's aura of an innovator, it is not hard to see that much of his work was clearly rooted in that of earlier, often nonacademic naturalists such as Zacharias. When Thienemann set out the research focus of his station in 1917 as "the interaction between the *Lebensraum* and its organisms in our inland lakes," this clearly echoed the program Zacharias developed twenty-five years earlier.[100] Thienemann's ecology had a holistic undertone similar to the work of Zacharias and it revolved around the same central metaphors, spatial strategies, methodologies, and ideas. Also in Thienemann's work terms such as *microcosm* and *Lebensgemeinschaft* were paramount.[101] Like Zacharias, he thought about lakes as wholes and he examined their characteristics particularly through comparative studies. Moreover, his lake typology clearly echoed Zacharias's distinction between nutrient-rich and nutrient-poor lakes. And Thienemann's ecological thinking in terms of production continued a tradition represented by Zacharias's work and that of contemporary fishery scientists.[102] All in all, there seems to have been more continuity between Thienemann's work and that of civic limnology than he was ready to acknowledge. Be that as it may, it is clear that it required a well-networked professional such as Thienemann to introduce the old civic interests in the academic world.

The science Zacharias had developed in the station of Plön was ecological in orientation. In many ways this orientation was comparable with the work Giard had set up in Wimereux. Both Giard and Zacharias fostered an interest in the long-term interaction of animals with their environment, and they worked out practices and concepts to tackle this interest using a permanent workplace in nature. Like Giard, Zacharias believed particular landscapes would be helpful in understanding larger problems of the life sciences, although the latter believed the (isolated) lake would be more promising in this respect than the coastline or the ocean. Furthermore there were some other conceptual divergences between the two station directors. Giard focused on the interaction of individuals with their environment (a field that would later be called "autecology"), whereas Zacharias was more interested in the question of how groups interacted with their milieu (or: "synecology"). This has to do with differences in the objects of study, sites of research, and intellectual background. Zacharias focused on plankton that lived in well-bordered lakes, and he approached this with a vision informed by Kiel ecology and Humboldtian holism. Giard, to the contrary, focused on invertebrate parasites, worked from the more heterogeneous landscape of Wimereux, and defended a perspective that drew on neo-Lamarckian theorizing and scientific reductionism. The main difference between the stations of Wimereux and Plön, however, seemed to have been one of social make-up. Giard used his station to develop an academic clan, and he spread his approach thanks to the careers of his students in various French universities. Zacharias's circle was of a very different kind, and his ideas largely traveled through lectures, popular works, handbooks, and fishery journals. Only thanks to Thienemann his type of science would eventually be brought to academia.

It might be clear that there existed an important intellectual transfer from the older civic traditions of limnological research to the modern academic discipline. The station of Plön would become a central place of this transfer, which was partially embodied by the succession of Zacharias by Thienemann. During Zacharias's own leadership the efforts to make limnology academically respectable and to use this new discipline to reform secondary school biology had largely failed. But the importance of the early history of his station lies elsewhere. Zacharias used the station in Plön to give the study of freshwater fauna in Germany visibility and a geographical center, the success of which can be measured from the critical acclaim and imitation abroad. He managed to gather the support of a heterogeneous group of stakeholders in order to fund the station, and by so doing gave limnology an early form of institutionalization. In the rhetoric in defense of his science, and in his way of practicing it, he developed new ways of looking at the landscape, bringing long-term interactions between the lake and its inhabitants to the center of study. These interests would eventually become academically respectable under Thienemann's directorship.

CHAPTER SIX

Rossitten

Moving Birds

Zoological stations introduced new scientific practices, fostered new social constellations, and provided these activities and bonds with a new professional status. The disciplinary framework in which they did so, however, was not that of zoology as a whole. The life sciences of around 1900 consisted of a myriad of (sub)disciplines that were only loosely affiliated. Field stations functioned, at least partially, within the logic of these (sub)disciplines. This is true for the seaside laboratories that brought about the academic rise of marine zoology, and it is equally true for the limnological stations set up to professionalize the study of freshwater organisms. It is even more evident for the stations devoted to maybe the most idiosyncratic of zoological disciplines: ornithology. While earlier biological stations had been used among others to unravel the intricacies of parasitism and lake microcosms, the ornithological observatory would particularly focus on another major problem in understanding animal life: migration.

Although focusing on different questions and functioning in a different context, we ought to understand the rise of the bird observatory as an integral part of the station movement. To begin with, the ornithologists clearly took inspiration from the successes of the marine and the limnological stations.[1] The fact that state support was given for such stations was rhetorically used to demand an equal financial treatment for ornithology.[2] But the inclusion of bird observatories in the station movement was more than merely rhetorical. Like marine zoologists and limnologists before them, ornithologists saw field stations as necessary tools to reform their research and put new types of questions on the agenda. Like indeed marine zoologists and limnologists, ornithologists claimed that stations were crucial for understanding living organisms, their behavior, and their interaction with their environment.

While the lake microcosm offers a well-delineated and geographically restricted object of research, this is far less the case for migrating birds. As a consequence, bird observatories generated practices of observation that differed in many ways from those of freshwater stations. At the same time there is at least one striking similarity. Both relied heavily on civic science and amateur work. Because of its geographical breadth, the study of bird migration needed the mobilization of a civic network that was even wider than that of the limnologists discussed in the previous chapter. This proved possible only because birds—culturally prominent and comparatively easy to observe—had a strong appeal outside the world of professional science. Ornithological stations tapped into this wide civic interest, which they further stimulated by mobilizing and coordinating increasing groups of amateurs. Despite the novelty and cultural resonance of these practices, ornithological stations have received only scant attention from historians. Because of an unceasing interest in the historiography of ornithology for professionalization, discipline building, and conceptual renewal, the focus of historians has particularly been on academic mandarins and influential museum curators.[3] This chapter, however, turns away from the prestigious centers of professional science and looks for (equally innovative) practices in more peripheral places of civic research.

This chapter focuses on one site of civic ornithology that would heavily influence ornithology at large. It delves into the history of the first permanent bird observatory in the world: the *Vogelwarte* in Rossitten, led by Johannes Thienemann.[4] Johannes had distant family ties with the limnologist August Thienemann (who figured so prominently in the last chapter), but the background of the two men could hardly have been more different.[5] Johannes was nineteen years older than August, and he was not an academic zoologist, but a theologically trained minister with a predilection for hunting and natural history. If anything, Johannes Thienemann was a civic ornithologist. The station he would direct was founded in 1901 in a remote area of then East Prussia as a private enterprise. Thienemann would lead the institution for the first thirty years of its existence. Despite its modest beginnings, the practices developed in Rossitten eventually became a model for those of subsequent bird observatories around the world. Without being a great theorizer or a discipline builder, Thienemann clearly left his mark.

This chapter will focus on the question of how Thienemann used the natural characteristics of the place in which he worked to develop his new approaches to the study and protection of living birds. At the same time, it will analyze how Thienemann continually redesigned the space he shared with the birds—materially, conceptually, and culturally. It will study this redesigning on the concrete level in Thienemann's daily work in the station and on the conceptual and the cultural level in his representations of Rossitten to a wider audience.

DETACHED BIRD LOVERS

The idea to set up a permanent ornithological station did not materialize out of thin air. Bird observatories have a history that dates back at least to the mid-nineteenth century. From the late 1830s onward, the influential Belgian statistician and astronomer Adolphe Quetelet set out to complement his survey of "periodical phenomena of the atmosphere" with those of the "periodical phenomena relating to plants, animals, and man."[6] The migration of birds was obviously one of the most prototypical of these phenomena and Quetelet used a European network of correspondents to monitor the departure and return of several bird species. His Royal Meteorological Observatory, situated in one of the rapidly expanding suburbs of Brussels, collected and processed the data.[7]

The establishment of (national) ornithological societies and (international) ornithological congresses in the latter quarter of the nineteenth century further stimulated the type of research initiated by Quetelet. At the inaugural meeting of the German Ornithological Society in 1875, research stations for the study of bird migration were already high on the agenda. In the aftermath of the First International Ornithological Congress in Vienna in 1884, an actual European network of temporary observation posts was developed. The following decade, this led to an almost endless collection of data. In the United Kingdom, for example, lighthouses and lightships were called in to gather as much information as possible. Yet, in Britain, as elsewhere in Europe, it was hard to actually link the data and nobody was really engaged in administering the observations. The disappointing results prompted European ornithologists to develop a new strategy. Instead of the loose network of temporary posts, the German Ornithological Society chose to establish one permanently staffed station in a favorable (and distinctively nonurban) place: the village of Rossitten.[8] By designating one particular center of observation, coordination, and accumulation, the Society opted for a new spatial strategy. As a source of inspiration it took the work of the painter-ornithologist Heinrich Gätke, who, since the late 1830s, had been observing migrating birds in complete isolation on the island of Helgoland. It was also Gätke's term of *Vogelwarte* [literally: bird observation post] that the German Ornithological Society would use for its first permanent ornithological station.[9]

The idea of setting up a station in Rossitten picked up on earlier initiatives. The first to dream of establishing a permanent bird observatory in the village had been Kurt Floericke, a merchant in both stuffed and living birds, who would subsequently have a career as a best-selling nature writer. In the 1890s he founded a modest *Verein vergnügter Vogelfreunde* [Society of happy bird lovers] in Rossitten, but his attempts to raise funds for a research station failed. Disappointed, he eventually decided to leave the region.[10] Yet Thienemann took over the project where Floericke had left it—even to the extent of marrying the latter's ex-wife. Through mobilizing the German Ornithological Society, Thienemann succeeded where Floericke had failed. The latter openly turned against the project and for many years he would remain one of the harshest public critics of Thienemann's station.[11]

The fact that Thienemann launched his initiative under the auspices of the German Ornithological Society—an independent learned body—was of consequence for the dynamic of the station. Like nineteenth-century ornithology as a whole, the German Ornithological Society had hardly taken part in the academization that characterized much of the late-nineteenth-century life sciences. Apart from some museum curators, it consisted predominantly of military men, nobility, teachers, pastors, and animal painters, who were not professionally engaged in research.[12] These men and the organizations in which they were involved acted as supplementary financiers of the station. Hence, the *Vogelwarte* was a "civic enterprise" institutionally unconnected to the rest of German zoology. The latter would change only in 1923, when the station was taken over by the Kaiser Wilhelm Society and (at least nominally) was made part of the organization's research network.

To be sure, the project of the observatory in Rossitten also triggered support from the academic world and state-employed scientists. Particularly influential in the development of Thienemann's project was Georg Rörig: former professor at the University of Königsberg, *Geheimrat* [or: Privy Councillor], researcher at the Imperial Health Office, and newly appointed director of the Biological Department at the Imperial Institute for Forestry and Agriculture. Rörig's connections with academia, state research institutes, and Prussian bureaucracy were certainly of importance in gaining political and practical support for the enterprise. Its scientific aura was furthermore enhanced by the support of academic zoologists—which, interestingly, consisted of largely the same group of people who had taken up the cause of Zacharias's limnological station. Karl Möbius, not only museum director but also influential Academy member, recommended the project to the Prussian Ministries of Culture and Agriculture, and so did the Leipzig professor and marine zoologist Carl Chun. Finally, the newly chosen president of the German Ornithological Society, Rudolf Blasius, also used his academic credentials as professor at the College of Technology at Braunsweig in support of the station. At the same time it is important to notice that Blasius' courses at the College concerned not ornithology, but hygiene. In the Germany of 1900, after all, the study of birds was an extra-academic occupation. The station in Rossitten, led by a pastor, would only continue this tradition.[13]

Not only because of its institutional set-up, but also because of its actual location, the Vogelwarte in Rossitten was largely detached from the (urban) centers of German science. East Prussia was a peripheral province. It belonged to the large areas east of the river Elbe—a river which present-day historians see as the "rural-urban axis" that divided the country.[14] It was the province with the lowest population density in the Reich and the highest percentage of the population working in agriculture. Politically, East Prussia was notoriously conservative, monarchist, and patriotic. Industry was poorly developed. The urban bourgeoisie (elsewhere the main supporters of scientific activity) was limited to Königsberg and Memel, which were the only sizeable cities in the area. In the rest of the province, social life was dominated by landed nobility (the so-called Junkers), by military men, agricultural societies, and pastors.[15] It was with these social groups that Thienemann had to get involved, if he wanted his project to succeed.

Even within the peripheral province of East Prussia, Rossitten was located at a peripheral place. It was situated on the so-called Kurish Spit: a peninsula at the Baltic coast 98 kilometers long, not broader than 4 kilometers at any point. In 1900, the small fishing villages of the Spit were still rather isolated and ethnically diverse (having a mixed Latvian, Kurish, Lithuanian, and German population). So-called wandering dunes dominated the landscape and made agriculture virtually impossible, except for a small patch of land near Rossitten. Almost everyone was engaged in the fisheries and contact with the outside world was limited. One could reach Rossitten with the steamboat services from the town of Cranz, but only in summer. The rest of the year one had to undertake travel by cart on roadless lands, which could take up to twelve hours for only 35 kilometers.[16]

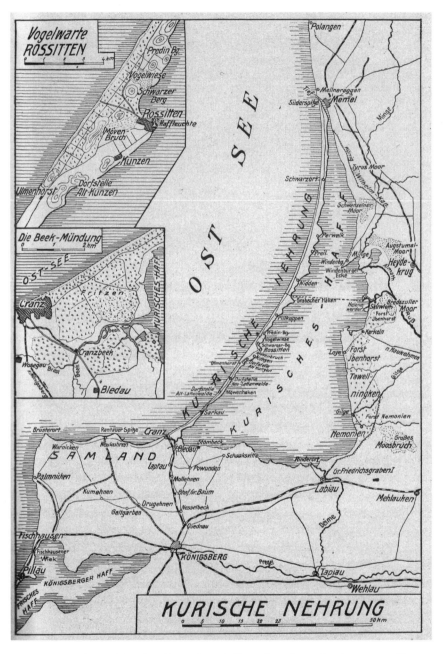

FIGURE 6.1 The Kurish Spit, with Rossitten halfway along the peninsula. From J. Thiene-mann, *Rossitten* (1930), 7.

Viewed from the urbanizing west, East Prussia was a place of tradition and nature, the so-called land of the dark woods. Yet, this was partially a myth. To begin with, agriculture had overtaken much of the forested areas, making the province in fact the least wooded in the whole Reich. And, despite what western city dwellers might have believed, modernity also showed its head in the rural societies east of the Elbe. Agriculture heavily intensified throughout the nineteenth century; transport was stimulated by the building of canals from the 1860s onward and the boom of the railroads after 1870. Several regions were Germanized (and, among others, the Latvian spoken on the Kurish Spit rapidly declined). Wilhelmine red brick schools, hospitals, and military barracks appeared in the wooden villages, and bureaucrats increasingly took over power from the Junkers. Königsberg saw its population increase from 73,000 in 1850 to 246,000 in 1900. In its surroundings, fashionable sea-resorts developed. Beginning in1895, a combined railroad and steamship network connected Königsberg via Cranz with the coastal villages of the Kurish Spit, which brought in tourists and landscape painters. The natural aura of the region was partially created by and for these new visitors. One of these external visitors was Thienemann, who traveled the 1000 kilometers from his native Thuringia for the first time in 1896.[17]

Thienemann had come to Rossitten for one reason: the village's established reputation for its abundance of birds. Rossitten was after all located on the path of important migratory routes for which the peninsula, according to Thienemann, served as a "natural direction indicator." He became convinced that the geography of the place made it an ideal spot for a permanent observatory. Because of the Spit's narrowness the whole area could be surveyed from the high dune crests. In the words of Thienemann, the region as a whole was "surveyable and controllable."[18] This (supposed) controllability of the Spit was rooted in its spatial isolation, located on a small peninsula essentially surrounded by large bodies of water. The same isolation that had made Zacharias's Grösser Plöner See into a microcosm could be used to turn the Kurish Spit into a natural laboratory to study bird migration. Next to controllable, the landscape was also diverse. As in the case of Wimereux and Plön this diversity was presented as crucial if one wanted to use one particular site to understand nature as a whole. Thienemann's study friend, the pastor and ornithologist Friedrich Lindner, used precisely this argument in recommending Rossitten as a place for the observatory to the Prussian government, stating that it offered "sea and lagoon beaches, forests, heath, wetland, sand deserts, wooded moorland, marshes, ponds, fields, meadows, gardens, wasteland, shrubs and reed" and all this on a few square kilometers.[19]

The Kurish Spit was not only a well-circumscribed (and diverse) area, but, as indicated, it was also isolated in the more popular sense of the word. Rossitten was a far-off place. Scientists such as Floericke and Thienemann presented this as an advantage. Unlike other sites of ornithological interest— such as the oft-visited seaside resort Helgoland—the cost of living was low in turn-of-the-century Rossitten and there was still hardly any beach tourism to disturb research.[20] If one wanted to study living birds for a longer period, Thienemann argued in 1910, one had to settle in unspoiled, free nature, just as some meteorologists had settled in the mountains.[21] Thienemann thus echoed an age-old theme that associated knowledge with isolation and that had been put to good use by earlier proponents of the station movement. In this rhetoric, the seclusion in Rossitten was nothing if not a scientific necessity.[22] At the same time, the loneliness of the scientist was presented as being agreeable. In his later work in particular Thienemann would highlight the romanticism of his isolated existence, emphasizing the originality of the area, the poetry of its wildness, and the primitive living conditions (far away from big-city life). The Kurish Spit was presented as both scientifically interesting and spiritually inspiring. "We become so small and humble in this majestic wild nature," he wrote in 1930.[23]

HOW TO OPTIMIZE NATURE'S ECONOMY

Bird migration was only one of the many subjects that the station in Rossitten was set out to explore. Its original building (the former studio of a landscape painter) might have been humble, but its tasks, as published in the bylaws of 1901, were ambitious. Next to migration, the *Vogelwarte* aimed to perform research in the areas of bird behavior, molt, and color change; it aimed to devise effective means of protecting birds and investigating their agricultural value; it aimed to establish a collection of ornithological specimens and procure material for scientific state institutes, and finally it aimed to serve as a center for popularizing ornithological knowledge.[24] Initially, all this work was to be performed by one man: the director and sole staff member of the station, Thienemann.

The research program of the station was implemented only gradually. Because his income was relatively low in the early years, Thienemann initially combined his research job with other occupations such as forestry accountant and agricultural rapporteur for the Department of Statistics in

Berlin. In his work for the station, then, it was particularly the study of bird protection that originally received most of Thienemann's attention. This interest, however, did not mean that he advocated preserving the nature of the Kurish Spit in its original state. Rather, Thienemann aimed at drastically changing the existing wildlife population in the area. As his involvement in forestry and agriculture accountancy might suggest, his view of protection was essentially a utilitarian one. He wanted to improve the household of nature by artificially increasing the number of "useful" animals and exterminating the so-called vermin.[25] This type of work might seem an odd project for a research station, but, as we have seen, it was certainly not uncommon in the station movement as a whole. After all, avoiding the depletion of fish stocks and increasing the production of ponds had been major objectives of late-nineteenth-century marine and limnological stations.

In his bird protection projects, Thienemann was particularly inspired by his friend, the officer, ornithologist, and popular writer Hans Freiherr von Berlepsch, who had been experimenting with nesting boxes and winter feeding since the 1880s. The latter devices had to protect birds that were perceived as useful for agriculture and forestry, but Berlepsch's influential book *Der Gesamte Vogelschutz* (1899) also argued for the purposeful extinction of their natural enemies.[26] It was a very similar utilitarianism that drove the research of another inspirer of the station in Rossitten, the aforementioned Georg Rörig. Around the turn of the century, Rörig set up large-scale projects of analyzing the stomach contents of birds in order to quantify the relative profit and damage that single species brought to German agriculture.[27] In the period before his station was founded, Thienemann had already involved himself in this project, by providing Rörig with hundreds of crows from the Kurish Spit.[28] The International Ornithological Conference in Paris in 1900 picked up on this type of work and recommended additional research. Subsequently, several countries set up bird stomach analyses, often financially supported by the government.[29] Also Thienemann presented the study of bird eating habits as one of the main tasks of his station, and defended this type of research with utilitarian arguments.

In the German bird protection movement, Von Berlepsch's and Rörig's instrumental and managerial ideas had been fairly prominent throughout the nineteenth century. In the early 1900s, however, these ideas would gradually be challenged by more ethically, aesthetically, and holistically inspired arguments. Birds, whether useful or not, increasingly gained an emotional appeal and nature was more and more romanticized as a place of harmony in which every animal had its task.[30] Thienemann, however, stuck to the old managerial school of thought. Only after his retirement would he adapt his viewpoint and state that every organism had its role to play in the "household of nature."[31] Yet his earlier protection (and extermination) projects seemed to suppose quite the contrary. And throughout the years these projects drastically changed the ecology of Thienemann's actual workplace.

One of the first managerial "protection" projects at the *Vogelwarte* was the attempt to artificially introduce hole-nesters to the Kurish Spit—a bird group that was, so far, largely absent from the region, but that was believed to play a positive role in the fight against the insect plagues that threatened Prussian forestry and agriculture. In order to ensure their survival Berlepschian nesting boxes were hung, "feeding trees" mounted, and "warmth apparatuses" installed. By numbering and regularly checking the nesting boxes, Thienemann tried to monitor his introduction experiment scientifically. To help the tits survive the winter, he would even feed them the carcasses of sparrows (birds with a far less positive reputation). Later, he would experiment with floating nesting boxes on the sea, in the hope of increasing the "production" of gull eggs in the region. Carnivores, on the other hand, were hunted down. Thienemann for instance encouraged the killing of foxes in the Rossitten region, in order to augment the hare population.[32]

In all these projects, Thienemann enjoyed the support of civic groups such as agricultural societies and hunting circles, whose utilitarian view of nature and economic agendas he shared, in whose journals he published, and at whose gatherings he lectured. Through the contacts with these groups, the "experiment" in Rossitten could be exported to the surrounding countryside. As early as 1902, Thienemann distributed 2000 Berlepschian nesting boxes throughout East Prussia as part of a project to contain the epidemics of oak processionaries.[33] Soon, a nesting box factory would open a branch in the area.[34] Such enterprises were not limited to Rossitten, for that matter. The governments of among others Germany, Belgium, and Hungary highly invested in the spread of nesting boxes on their national territory. In Hungary, the Royal Ornithological Office even owned its own (highly productive) nesting box and bird food factory.[35]

RINGS AND EXPERIMENTS

The focus of Thienemann's interest shifted relatively quickly. His tests with Berlepschian nesting boxes and feeding trees gradually gave place to an ambitious study of bird migration. This was not such a drastic shift as it might seem. Thienemann after all started to believe that in order to successfully protect migratory birds (or rather to manage them as a natural resource) one had to trace their habits and whereabouts. This interest can be framed within the growing popularity of migration studies that characterized the turn-of-the-century life sciences as a whole. Since the 1870s, the fishery biologist Friedrich Heincke, for example, had set up extensive surveys on the migration of herrings—a study for which there were also clear economic motives. Despite strong analogies in program and set-up, however, the methodology of Heincke's and Thienemann's projects drastically differed. The first tried to unravel the mystery of herring migration by using novel biometric techniques for discerning the herring "races" which he believed formed different migratory groups. Thienemann, for his part, tried to track the movement of individual animals, and he did so by employing another recently developed technique: banding.[36]

Thienemann started his so-called bird-banding experiment as early as 1903. He would continue the project until his retirement and it would earn his station an international reputation.[37] The idea of banding birds was no complete innovation, but it was only taken up systematically by the Danish schoolteacher Hans Mortensen in 1899. When Thienemann heard about the new research tool, he was immediately enthused. "Before," he recollected in 1922, "the research ended when the bird had flown over. Now it continued."[38] People who observed ringed birds (dead or occasionally alive) could report them to Thienemann and return the ring to him. Using a system of numbers printed on the ring, Thienemann was able to identify the individual bird and the route it had taken. The system included only a relatively small sample of birds but, nonetheless, early experiments had an immediate effect. Within just a few weeks of the first birds being released, ringed specimens were already being reported. Thienemann himself was surprised by the success. "I really did not assume I would have results so quickly," he wrote to fellow-ornithologist Herman Schalow in November 1903.[39]

Despite the successes with banding, Thienemann also tried out other techniques. Ringed birds could not be recognized from far away so experiments were devised using specimens marked with colorful paint. These alternative techniques were quickly abandoned, however. The painted birds made a "pathetic impression" according to Thienemann and the paint came off after just a few months. Yet Thienemann's most important objection was that painted birds were no longer "natural" research objects. While rings were believed not to take birds beyond the realm of nature, paint actually turned them into "objects of art."[40] As in other stations, it was "unspoiled" nature that constituted the subject of study.

While painting birds was given up quickly, other observational practices continued alongside banding. Thienemann, among others, kept diaries, in which he wrote down which species he had seen passing, their approximate numbers, and how many he had actually caught. To this he added detailed records of weather conditions, noting down the barometer readings, air temperature, wind direction, rainfall, and daily hours of sun. The hope was, again, to link the behavior of animals, in this case their migration, with the conditions of the environment in which they lived. This proved rather difficult, however, and, although Thienemann continued these types of observation, it was particularly the project of banding that eventually became most prominent in his yearly reports.

The system of banding was developed for different kinds of birds, but the bulk of the experiments in Rossitten focused on a limited number of species. The choice was a practical one and intimately linked with the possibilities the concrete workplace offered. At first, Thienemann mostly worked with hooded crows. The Kurish Spit after all had a significant number of "crow catchers," who caught thousands of crows annually for human consumption. It was easy and cheap to procure living specimens from them, so that Thienemann had an auxiliary workforce immediately to hand. Moreover, the chance of recovering the rings was fairly high since—in the words of Thienemann—"no hunter leaves a crow unshot when it flies over within shooting distance."[41] Thienemann also discovered rather quickly that the Kurish Spit constituted a bottleneck in the migration of hooded crows. In his later work, he would write that the Spit is nothing less than "the place where one can feel the pulse of this great migratory organism; it is the touchstone of what goes on in this vast hustle and bustle."[42] Rossitten, so he claimed, was one of the only places where one could survey large populations, while working at one and the same spot. In his choice of his workplace (and his conceptualization of it) Thienemann showed his aptitude in practices of place.

Gradually, Thienemann extended the number of reference species with which he worked. In 1905, he began large-scale ringing of black-headed gulls. These had similar advantages as crows. The gulls were very common in Rossitten; using nets it was easy to catch their young when they swam on the water, and adults could be obtained from local crow-catchers. Often hunted, the rings came back in rather high percentages.[43] In 1906, then, Thienemann also started to include white storks in his research. These were far less hunted—at least in Europe—but they were big and conspicuous, cherished by the local population, and they had the advantage of being able

FIGURE 6.2 A Rossitten "crow-catcher" at work. (Courtesy of the Max Planck Institute for Ornithology, Radolfzell)

to carry rather heavy rings containing a lot of information. Finally, the stork was the prototypical migratory bird flying south. This bird would soon become the unofficial emblem of the station.[44]

The successes with banding crows, gulls, and storks did not quash all criticism of Thienemann's project. Critics argued that the aluminum harmed the birds' legs and that the already declining groups of migratory birds would be killed more often in order to recover the rings. It was painful for Thienemann that this criticism was heard particularly in a civic world he himself tried to master: the world of nature writing with its mixed interests in conservation, hunting, and the popularization of science. Kurt Floericke—by then editor of the influential popular science journal *Kosmos*—led the attempts to discredit Thienemann. Floericke described bird banding as "vain scientific humbug" and predicted "mass murdering of storks." In these attacks, Floericke's allies included the popular nature writer and "poet of the heath" Hermann Löns. In his best-selling *Mümmelman: Ein Tierbuch* [Hare: An animal book], Löns took up a conversation between two crows highly critical of the *Vogelwarte*. Thienemann, so the talking birds believed, ordered "that crows everywhere be shot and their feet sent to him, in the name of science." Elsewhere, Löns criticized Thienemann for his presumed "collection mania," thus associating him with out-of-date natural history practices rather than the state-of-the-art aura the director of the *Vogelwarte* was seeking for himself. As an "animal lover," Löns stressed that he believed "one single stork to be more beautiful and more useful than all ornithologists taken together."[45] His criticism clearly reflected the growing role that an "emotional" attachment to animals started to play in the world of German, and indeed Western, nature protection.[46]

Yet, the bird-banding "experiment" outlived the criticism and Thienemann continued to increase its scale. This expansion went hand-in-hand with a change of strategy. Thienemann's idea to start banding storks was indicative in this respect. After all, storks were rarely, if ever, seen in Rossitten.[47] Thienemann could study them only by distributing rings to amateur birdbanders in neighboring territories where the species was more common. He made sure everyone knew that Rossitten delivered rings "free of charge and postage paid" to all amateurs.[48] In this way, he managed to engage a wide civic network. His strategy echoed that of the large-scale surveys of turn-of-the-century British plant ecology, which also basically relied on the recording work of large groups of amateurs.[49] Closer to home, similar tactics were used by the Danzig museum curator Hugo Conwentz, who surveyed plant and animal species by handing out circulars and questionnaires to the local citizens and public officials.[50] In the Rossitten case the success of this mobilizing strategy was immediate, and soon the birds ringed outside Rossitten easily outnumbered those ringed at the station itself. Thienemann recruited state foresters, zoology professors, and wealthy landlords throughout the country to be responsible for ringing. He prided himself on the fact that he invested a lot of time in maintaining this network, because he believed only personal contacts guaranteed the trustworthiness of the information.[51] It enabled him to observe nature far beyond his place of residence. Rossitten remained the center where all the data were gathered, but only some of the birds represented by crosses on Thienemann's maps actually flew over his station. To be sure, Thienemann continued to observe individual specimens in the field, but he was only a virtual observer of the majority of the birds he monitored.

In addition to a network of volunteers to ring birds, Thienemann obviously needed people to report the finds and send him back the rings that were found. Therefore, he invested a great deal in publicizing his project throughout the country. He lectured to all kinds of scientific and agricultural societies, showing slides and eventually films; he organized summer courses and published his work in hunting journals and natural history magazines. Those who reported finds were rewarded by the fact that Thienemann included their names in his lengthy annual reports. Thanks to these, we have a fairly good idea of the diversity of the people engaged in the project. Noblemen, soldiers, and schoolteachers often appeared in the lists, as well as foresters and hunters, state officials, students, physicians, and pharmacists. Professional ornithologists often acted as intermediaries between Thienemann and the fieldworkers.[52] They also proved of great value in sending him articles from local newspapers, reporting on finds of ringed birds.

Obviously, as Thienemann himself observed, his "winged research objects flew, without taking political borders into account."[53] For finds outside Germany, German expatriates initially played a role, but very soon volunteers of other nationalities also became involved in the project. For reports of finds in the Middle East and Africa Thienemann relied on colonial networks. Often, the natives made the original find (by shooting a stork, for example), but it was only the colonials who reported it. Missionaries, embassy staff, and colonial scientists played key roles as go-betweens. In particular, monitoring the migration of storks from Scandinavia to South Africa—the best-known project of the station—relied in large part on a cultural milieu shaped by Western imperialism.[54]

The First World War disturbed international collaboration for a while, and in its aftermath Thienemann's station would become an even more peripheral place than it already was. The establishment of the "Polish Corridor" cut off East Prussia from the rest of the Reich, while the foundation of the Lithuanian state brought the national border very close to the *Vogelwarte*, cutting the Kurish Spit in two.[55] These developments led to increasing nationalist resentment in the region, but Thienemann was not carried away and was quick to re-establish the old international contacts. "Tiny aluminum rings," so he wrote in 1919, "are able to further relations between nations." As a result of the extended networks involved, Thienemann argued that his science was nothing if not internationalist. It was "the common labor of the races" and "a work of peace."[56]

At the same time, however, Thienemann stressed the role that his work could play in the growing *Heimatliebe* [love for the homeland], like indeed Zacharias's plankton studies had done. In Thienemann's own view, his lectures to local youth and his popular courses not only popularized science, but also strengthened the general sense of the *Heimat* by alerting people to the beauty of local nature. This mixture of science, conservationism, and the promotion of *Heimatliebe* was by no means exceptional. It was wide-spread in the German nature protection movement of the early twentieth century.[57] More striking is that the theme of the *Heimat* found its way into Thienemann's actual studies of bird migration. Was it not the sense of the *Heimat* or even straightforward *Heimatliebe* that guided migratory birds back to their original nests? These birds might have been international travel-ers, but they were also strongly locally rooted. And Thienemann perceived this rootedness as natural and ethically good—with regard to both birds *and* people.[58] Eventually, these ideas would, in the 1930s, find a particular echo among the National Socialists—albeit somewhat despite Thienemann's own interpretation. To stress the connection of East Prussia with the rest of the Reich, the Nazis attempted to attract (western) German tourists to the re-gion, among others by referring to Thienemann's bird research. On stamps of the period one could see migratory birds flying over a map of the region proclaiming: "We travel to East Prussia every year, and you?"[59]

When referring to his work of bird banding, Thienemann systematically spoke of an "experiment." Again, he followed strategies developed earlier in the station movement. Thienemann chose the term *experiment* for its rhetor-ical power, suggesting a laboratory-like precision. Writing about the "band-ing experiment," he stated: "All hypotheses end there; facts are created. That is what makes experiments so attractive."[60] Obviously, Thienemann's bird banding was not an experiment in the sense it received in nineteenth-century physiology.[61] It was performed in "real nature" and manipulation was kept to a minimum (not to interfere with the birds' natural behavior). Thienemann, however, used the term to indicate that his work delivered precise, controllable facts, unlike earlier, more hypothetical migration stud-ies. It was largely a semantic trick to associate his research with the prestige of the laboratory sciences. In this way it resembled Giard's conception of nature as a laboratory, or Zacharias's representation of the lake as a research aquarium. Their intellectual strategy, as indeed that of many leaders of the station movement, consisted of both claiming part of the aura of laboratory science, while at the same time proudly defending their outdoor identity.

After having worked out the banding system, Thienemann tried to ex-pand his experimental approach to other aspects of ornithology. Among other things, he developed a set-up to study the speed of the bird's flight. For this, Thienemann would retreat to Ulmenhorst—an annex of the station, opened in 1908, in a "world-forgotten place" some seven kilometers outside Rossitten. The Spit was at its most narrow here and migrating birds tended to fly over it in straight lines. Again, Thienemann demonstrated his skill in developing scientific practices of place. He would set up two observation posts connected by field telephones. The task of the observer at the first post was to spot a bird flying over, to start a chronometer, and to contact the second observer over the telephone. The latter would give a telephone signal when the bird flew over the second post. Using some rudimentary measuring instruments, it was possible to take the speed and direction of the wind into account. The set-up was far from spectacular, but the results were generally seen as innovative. At least, they were thought to be more precise than any estimates so far.[62]

Other field "experiments" that Thienemann developed in Ulmenhorst concerned the altitude of flight. This issue had been heavily debated ever since Gätke started to publish on the subject in the 1840s, developing the theory that birds generally flew at a height of several thousand meters.[63] Thienemann tried to falsify these hypotheses. He did so particularly in interaction with his friend, the lieutenant, amateur ornithologist, hunter, and popular writer Friedrich von Lucanus. Thanks to his military background, Lucanus was able to carry out various observations using hot air balloons from the Prussian army. Among other things, he released birds at different altitudes, noticing that they would only leave the immediate vicinity of the balloon once land was in sight. Furthermore, he suspended stuffed specimens under the balloon to investigate up to what altitude they were actually perceivable from the ground.[64] Thienemann was inspired by Lucanus, but his station did not have a balloon at its disposal. The station's director therefore set up improvised experiments with stuffed birds suspended under kites or placed on high poles in order to estimate the altitude of passing birds in its environment. He admitted that such experiments were somewhat primitive, but valued them above earlier estimates with no reference points whatsoever.[65] Thienemann hoped modern technology—especially airplanes—would eventually enable the scientists "to transcend time and space" and thus solve the question in a more precise manner.[66]

Thanks to occasional collaboration with the gliding pioneer and popular hero Ferdinand Schulz, Thienemann was at least able to take a step in this direction. Rossitten, with its long dune strips and particular upward currents, proved exceptionally suitable not only for bird-watching but also for gliding. In the early 1920s Schulz set several world records in Rossitten and (from 1925 onward) the village would be home to the world's first gliding society.[67] Gliders and ornithologists were thus partially attracted by the same type of landscape and Thienemann was very proud to incorporate some observations by the renowned Schulz in his later work relating to both the altitude and speed of the bird's flight.[68]

As indicated, Thienemann's experiments on the speed and altitude of flight scarcely entailed any manipulation of variables by the scientist. In the late 1920s, however, he developed experiments that incorporated this element of prototypical laboratory culture. The question Thienemann concentrated on during this period was how migratory birds oriented themselves. He wanted to know whether young birds learn migration routes by accompanying their parents or whether they follow a genetically inherited instinct. To actually test this, Thienemann devised his so-called release research. From 1926 onward, young storks that had been reared in Rossitten were released, but only after the older birds had already left. To monitor this precious group as closely as possible, Thienemann again resorted to modern technology. He announced the release of the storks on local and national radio stations, thus triggering increased awareness among the population. This strategy worked and Thienemann received particularly detailed reports. Reports from outside Germany were fewer in number, but enough to gain a general picture. Overall, the birds largely seemed to follow the normal route, but eventually turned west, whereas most of the older birds continued south. The results, in other words, were somewhat mixed and Thienemann—close to his retirement—was not able to settle the orientation question completely.[69] It still preoccupies scientists today.

THE HAT OF THE HUNTER

In the thirty years he worked at Rossitten, Thienemann clearly incorporated some conceptual and material elements of modern laboratory science in the field. He conceptualized the space in which he worked in such a way that he could isolate the phenomena he wanted to study. He introduced new measuring techniques and went on to manipulate the organisms with which he worked. Yet, in other aspects his attitude was far from the ideal of the lab. Like several other contemporary station directors, he would explicitly stress that modern zoology had become too laboratory-oriented and that it "urgently needed more fieldwork, outside in free nature."[70] Thienemann himself had no contact whatsoever with the new zoological laboratories that had been erected at the universities. He was an autodidact, who would not receive his degree in zoology until he reached the age of forty-five.[71] Almost immediately he was appointed professor at the University of Königsberg, but he remained very much an outsider there and would never become a typical academic.[72] When appearing in public his clothing was not unlike that of the woodsman—wearing high boots and a hunter's hat, very unlike the more formal costume of the average metropolitan scientist.

FIGURE 6.3 Johannes Thienemann equipped (and dressed) as a hunter-naturalist. (Courtesy of the Max Planck Institute for Ornithology, Radolfzell)

Thienemann was a late representative of a nonacademic, rural tradition of nature study, in which the parsonage rather than the university was the center of knowledge. In nineteenth-century Germany, evangelical ministers (including Thienemann's direct ancestors) had played an important role in the development of this nature study. The work of these pastors had been practical, rather than theoretical, closely connected to hunting and agronomy, teaching and popularization—as indeed the work of Thienemann himself. Although he would break away from the overt natural theology and anthropomorphism of nineteenth-century evangelical nature writing, Thienemann could not hide his roots.[73] One could see an evangelical influence in the striking absence of evolutionary theory in his work, in his talk of natural harmony, and in the fact that the Ulmenhorst station carried

an eye-catching sign "In honor of God and his Nature." In most of his ornithological work, however, religion was not particularly prominent and
its continuity with the evangelical tradition has rather to be sought in its
popular, practical, and nonurban outlook.

Thienemann functioned in a network that differed significantly from
that of the average academic laboratory scientist. Whereas the latter conducted his science primarily in their "intramural counterparts" inhabited
by a homogeneous group of experts, Thienemann worked in the "open
space" of the field and collaborated to a significant extent with nonscientists.[74] His network ranged from crow-catchers to state officials and was
of a mixed, largely nonscientific character. The same is in fact true of his
activities. In addition to his scientific work—and intrinsically connected to
it—Thienemann hunted, reared fowl, lectured on agriculture, acted as an
accountant, and guided influential military men and politicians around the
Kurish Spit. When the young Niko Tinbergen (the later ethologist) visited
the *Vogelwarte* in 1925, he noticed that it was visiting officials rather than
scientists who took most of Thienemann's time. It was such men, after all,
who not only dominated East-Prussian politics, but also its civic scientific
organizations.[75]

Thienemann certainly used his heterogeneous network and multiple
identities to his advantage, but these also created their particular kinds of
problem. From his early days at the Spit, Thienemann was at odds with
the local dune inspector, who charged him with game theft and who even
conducted a domiciliary visit to his house. In a letter to Berlin, Thienemann
stressed how this undermined his prestige with the local population. Later,
several conflicts ensued between Thienemann and Rossitten's chief state
forester. The two continuously competed over the municipal hunting lease
and their relationship quickly became sour. On several occasions the state
forester openly turned against Thienemann, accusing him of unlawfully
collecting eggs and entering certain forbidden areas. Eventually this led
to Thienemann being summoned by his superiors to Berlin.[76] The Kurish
Spit may have been easily controllable in a geographical sense, but actual control of the region also implied power in the social domain. Here,
Thienemann's mixed profile was sometimes more of a hindrance than an
advantage.

Nonetheless, Thienemann continued to cultivate his image of a hunter
and an outdoorsman—rather than that of a scientist's scientist. This left its
mark on his published work. Unlike academic scientists, he only rarely referred to the state of the art in his research field, stressing that he had no

interest in "armchair learning."[77] His publications were to give a first-hand account of his personal work in the field. This was translated into long descriptive lists (of ring finds, for example), but also into all kinds of anecdotes in his annual reports. In the reports of 1923 and 1924, for example, he talked at length about a stork found in Spain with a "Negro arrow" in its back, which Thienemann was able to trace back to the region of Sudan and East Africa.[78] Elsewhere, he extensively described how the Bulgarian press mistook the number on the ring of an eagle for its date of birth, concluding that it was 726 years old.[79] This kind of story was used to spice up the long inventories of "facts" that Thienemann had gathered. The same mechanism can be found in the small museum he set up at the *Vogelwarte*. Alongside rows of stuffed specimens, which inventoried the bird species of Rossitten, one could admire a rarity cabinet, displaying the objects relating to Thienemann's anecdotes: the Negro arrow in question, next to all kinds of ringed bird legs and the rubbish found in stork nests, from ladies' gloves to a potato.[80] To Thienemann's mind, these elements made his research tangible to both the general public and the popular press—two groups he desperately needed in order to actually perform his work.

Much of Thienemann's anecdotes were written in a humorous, somewhat masculine tone, not uncommon in the hunting literature of the time.[81] This is especially true of his later best-selling books, in which he portrayed the roughness of the Kurish Spit and the hardships of working there. Thienemann made clear that he disliked "sentimental" animal protectors and he wrote sympathetically about schnapps-drinking, wife-beating fishermen or crow-catchers who killed crows with a single bite.[82] In his most popular book, *Rossitten: Drei Jahrzehnte auf der kurischen Nehrung* [Rossitten: Three decades on the Kurish Spit], he described the so-called Kentucky Club in detail, which he formed at Rossitten and which was basically a group of gun-lovers. To become a member of the club, one had to shoot a pig using a single bullet; to become its president one needed to have received at least one shot wound and then healed without medical assistance. Members smoked pipes (not effeminate cigarettes) and paid with cartridges (not pernicious money). They aimed at moral and bodily discipline through a tough way of life: "The Boys—as the members are called—have to scrupulously avoid everything effeminate and prissy."[83] Thienemann said all this in jest, but of course nothing is more serious than humor. The self-image he maintained was strongly associated with the impression he gave of his scientific work. Observing birds in Rossitten was nothing for spoiled city dwellers, but rather for fearless, manly, and practically oriented huntsmen like himself.

FIGURE 6.4 A corner in Thienemann's anecdotal station museum. In the front drawer: a collection of rubbish found in a stork's nest. (Courtesy of the Max Planck Institute for Ornithology, Radolfzell)

Typically, when Tinbergen showed up at the *Vogelwarte,* Thienemann took him on hunting trips rather than initiate him in the scientific particularities of bird migration study.[84]

Thienemann's publications spread the image of the unspoiled and adventurous Rossitten throughout Germany. His books were reprinted various times, editions for children appeared, and postcards of the region were distributed.[85] In 1928, Thienemann even issued a four-reel movie called *Im Lande des Vogelzuges* [In the land of the bird migration], shot by the Berlin company Hubert Schubert and directed by Thienemann himself. The film closely followed the story of Thienemann's book on the same subject, but translated it for an even broader audience. It stressed the originality and unspoiled character of the Kurish Spit and its population with shots of traditional fishing boats and horse-drawn carriages, crow-catchers, and weather-beaten fishermen. The scientific story was told in a similar anecdotal fashion. Shots of bird-ringing were interspersed with images of bird behavior (such as storks eating), accompanied by casual commentary

("Surely a hungry company . . ."). Animated maps alternated with shots featuring animals with human names, such as the station's eagle owl, Hannah.[86] Again, Thienemann proved himself an apt popularizer.

Through his popular books and films, Thienemann triggered a nationwide attention, which resulted in a growing number of amateur ornithologists and nature-loving tourists who were drawn to Rossitten. By the year of Thienemann's death, the small village had seven hotels, as well as numerous bed and breakfasts. Tourist brochures praised the broad beaches and the quiet of the traditional, car-free village.[87] Yet, the growing number of visitors, attracted by images of unspoiled nature, drastically changed the place itself. Thienemann's own attitude toward this invasion was ambiguous. It was good for the reputation of the *Vogelwarte* (and thus for the migration experiment) and it enabled him to present his station "as an outpost of culture in the middle of the 'desert.'" At the same time, however, he openly lamented rising prices and the disappearing wildness of the area.[88] In his books he therefore preferred to write about the good old days.

Thienemann's reaction to the tourist development of Rossitten reflects the ambiguity typical of his nature image. This image borrowed from the two prevalent nineteenth-century traditions in Western thinking about nature, which the historian Donald Worster has described as "arcadianism" and "imperialism." The first is nostalgic and strives for a life in close harmony with nature, the second aims at a dominion of nature (among other means through science).[89] Thienemann strove for both. Authentic Rossitten, untainted by tourism, was to be both civilized *and* preserved. With many scientists of his time, Thienemann shared an "enlightened" civilizing mission and a love of technology, but he was also a cultural pessimist, criticizing the artificiality of much of modern culture and longing for a refuge of authenticity. Rossitten in many ways offered such a refuge and Thienemann would romanticize its original state. As a seventy-year-old, he described how the unspoiled landscape had touched him when he first visited the place. "On the Kurish Spit," he wrote, "I felt the sense of the magnificent, the new and the undiscovered and I came to know life in a form untouched by the outside world, so full of romance, so unspoiled, rough, rugged and primitive."[90] If one wanted to understand life in its "untouched" form, which was the central aim of many prominent figures of the station movement, than one had to settle in such a place.

Obviously, the ambiguous combination of an arcadian and an imperial nature image can be found in the work of several scientists. Biological field stations (with their mixed status as places of scientific conquest and retreat) offer a good breeding ground for such ambiguities. These are, among others, perceivable in the publications of Giard and Zacharias. Yet, in Thienemann's writing they are more articulate and thematized. The peripheral geography of Rossitten made it into a "wilder" place than Wimereux or Plön, which implied that there was more wildness to be conquered and/or to be cherished.

In letters to Berlin friends, Thienemann paraded his self-chosen loneliness in backward East Prussia, describing how he was "living in the dunes, with my dog, far from all human culture."[91] Yet, at the same time, the self-proclaimed woodsman did not always cope well with the hardships of the place where he lived. In 1910 he requested the permission of the German Ornithological Society to spend his winters in the provincial town of Cranz instead of the dark, isolated village of Rossitten. A few years later, he even made the metropolitan city of Königsberg his winter quarters.[92] In the last article he was to publish, Thienemann wrote that anxiety often came over him when, after a visit to Berlin, he traveled back to the Kurish Spit. But eventually, so he suggested, his love of nature and authenticity would always prevail. He stated that swarms of migratory birds observed from a dune crest always erased his craving for the city and made him feel the urge to shout: "O Berlin, I do not begrudge you your clean people, your champagne and oysters, what I have here is something you cannot offer."[93] Rossitten was the anti-Berlin. The romanticism of its wildness, as indeed that of so many other places, was constructed in response to an increasingly dominant urban culture. As such, it could exist only together with its counter-image.

INTO ACADEMIA (BIS): SCHÜZ'S ROSSITTEN

While Thienemann was publishing his most popular, nostalgic work, the directorship of the station was taken over by a scientist who was almost forty years his junior. His name was Ernst Schüz. Although Schüz would clearly build on Thienemann's work, he differed from his predecessor in many ways. Unlike Thienemann, Schüz was trained as a professional biologist and, as a former student of the omnipresent Erwin Stresemann, he was well acquainted with the ins and outs of the profession. Schüz may well have liked many aspects of Thienemann's work, but he clearly wanted to set matters on a new footing once he joined the staff in 1929.

Schüz founded a new journal, redesigned the *Vogelwarte* museum, and introduced a new vocabulary, more laden with jargon. The term "ecology," for instance, would appear for the first time in the station's publications.[94]

Schüz also strengthened the station's laboratory focus. In one of his annual reports he defended this policy as "simply a duty with regard to the scientific prestige of the station."[95] He launched histological and physiological research projects, particularly focusing on the effects of hormones on bird migration. Microscopes and microtomes were bought, as well as extraction appliances, thermostats, and a vacuum distillation machine.[96] The new research focus was partially developed in coordination with the other bird observatories in the Reich. In the mid-1930s government officials had created a "working group" that institutionally united Schüz's station with the observatories at the biological station in Helgoland (which had been set up in 1910) and the ornithological annex of the university of Greifswald at the Hiddensee (founded in 1936).[97]

It was not only the material and institutional situation of the *Vogelwarte* that changed. The political milieu in which the station operated also changed drastically. Once the National Socialists were in charge, the station depended directly on the politically dominated Emergency Association of German Science [*Notgemeinschaft der deutsche Wissenschaft*]. Within the Kaiser Wilhelm Society it was known that the leadership of the Emergency Association was inclined to help "the hard-pressed eastern province of the Vogelwarte, because of political reasons."[98] In official documents, Schüz and his Berlin director Oskar Heinroth therefore started to stress the strategic importance of ornithological fieldwork at the Kurish Spit for *Deutschtum* [or: Germanness]—since the area was also regularly visited by Lithuanian scientists from across the border.[99] Furthermore, the discourse about the role of the *Vogelwarte* in shaping a general feeling of *Heimat* evolved in a National Socialist direction. In an article dated 1934, Schüz referred to the duty of his station when he rhetorically asked: "Will this duty ever find more approval than today, a time when *Blut und Boden*—which in the words of the *Führer* are the building blocks of our *Volk*—are measured using another, better standard than ever before?"[100] Yet, despite such rhetoric, the actual scientific work at the *Vogelwarte* seemed not to have been influenced too much by the change of regime. A project devised by Heinrich Himmler in 1942 to use the storks from Rossitten to spread political propaganda was successfully countered by Schüz.[101]

Thienemann was not to live to witness Himmler's plans, nor the takeover of the station by the Soviets in 1945. During his lifetime, however, Rossitten had already changed drastically. The place where pioneers such as Floericke made a few isolated observations in the 1890s was a far cry from the place that hosted Schüz's well-equipped station in the 1930s. Roads had been constructed, bakeries had opened, and the foxtrot had replaced the traditional dances. Nesting boxes had been put up, new species had been introduced (while other species had become rare), a museum had been built and redesigned several times, and field annexes had been put up. In 1901, the station's equipment largely consisted of binoculars and a gun; forty years later, all kinds of laboratory apparatuses had been introduced. In the scientific world, the name of the village had become internationally known. It spread all over the world—printed on metal rings. Ornithologists knew that Rossitten was a "bottleneck" on the routes of migratory birds, as well as a center for the accumulation of ornithological knowledge. The station hosted only a small staff but it welcomed many visiting scientists and mobilized amateurs on three continents. In Germany, Rossitten became an icon of unspoiled wilderness, attracting tens of thousands of tourists each year.

Thienemann stood in the midst of these changes. He introduced a new kind of ornithology, not through novel theoretical concepts, but through specific practices of place and through mobilizing a largely civic network. After Thienemann's appointment at the university of Königsberg and Schüz's selection as director of the station, academic culture made its way to Rossitten (as it had in Plön) but only partially so. The older nonacademic traditions and networks lived on and remained crucial for the station's functioning into the 1930s and beyond. Later, Thienemann's interest would be recognized as being *ecological*, but he never used this term himself.[102] Thienemann might have been inspired by an experimentalist tradition, but above all he was a hunter-naturalist, who sought poetry as much as precision. Thienemann turned the Kurish Spit into a laboratory of nature and initiated the construction of a scientific infrastructure at the peninsula, where he introduced, caught, ringed, chronometered, bred, and shot birds. Additionally—as an integral part of his science—he put a great deal of effort into making his workplace known among fellow scientists, amateurs, and the general public. While doing so, he subsequently presented Rossitten as an outpost of civilization, a center of modern science, and a place of wilderness and masculinity. In all these ways, a small village in East Prussia was simultaneously transformed into a genuine scientific workplace, a conceptual tool, and a cultural icon.

The intellectual strategy of Thienemann had been comparable with that of several turn-of-the-century representatives of the station movement. It aimed at studying animal behavior in interaction with the environment by making the landscape into a research tool. Yet, his social and spatial strategies differed from those of many of his colleagues from other biological disciplines. He was much more ambitious with regard to the surface of the area he explored and the number of the (amateur) collaborators he involved. Choosing moving birds as a research topic necessitated a large-scale approach for both. Because of his large-scale projects, Thienemann became a pioneer in making ornithology into what David Allen has described as a "permanent cooperative enquiry."[103] In many ways, he was well suited for this job. Thanks to ingenious practices of place he managed to observe a large geographical area from the well-selected site in which he had set up his station. Thanks to his civic networks and popular books, film production and museum exhibits, he was able to mobilize a large group of people and to give renown to his work.

All this did almost immediately lead to scientific following. Particularly Thienemann's banding proved influential and inspired similar projects in other parts of Germany, and the rest of Europe from the 1910s onward.[104] These projects were not always set up from field stations, but this would be increasingly the case in the following decades. Ornithological stations were founded among others in Helgoland (Germany, 1910), Liboch (Austria, 1914), Sempach (Switzerland, 1924), Garda (Italy, 1929), Neschwitz (Germany, 1930), Texel (the Netherlands, 1931), and Skokholm (United Kingdom, 1933).[105] Like Rossitten these places often mixed issues of protection with research. By the time Schüz took over the leadership of the *Vogelwarte* in Rossitten, a European network was in place with established practices and guiding principles inspired by his predecessor. Partially, this tradition would be drawn into academic culture, but without losing a scientific, social, and spatial continuity with the past.

CHAPTER SEVEN

Brussels

Fieldwork in a Metropolitan Museum

In the decades around 1900, the station movement played an important role in the geographical dispersal and decentralization of the community of life scientists. For most of the nineteenth century, zoologists had institutionally been concentrated in a limited number of cities; now, they found a home in coastal villages, provincial lakeside towns, and isolated observatories in the backwoods. This scattering was the result of the intellectual program that underlay the station movement, and that promoted a type of knowledge that was local and based on the study of living nature. As such, station zoologists often contrasted their own work with that of "museum zoology," which was rhetorically presented as a project of urban sedentary naturalists interested only in classifying dead organisms that were detached from their habitat. In this rhetoric, the metropolitan natural history museum was the contrary of the field station.

In practice, the contrast between stations and museums was not as clear-cut as it was in the self-serving caricatures of some of the scientists involved. Much of the scientific practice in stations entailed killing organisms, or, at least, detaching them from their natural habitat. Like their colleagues in museums, station zoologists set up a lot of survey work and created reference collections. In some cases, like the museum curators, they displayed their specimens for a nonspecialist audience. Stations, in short, integrated a lot of museum manners. At the same time, there was also a transfer of practices in the other direction. Unlike what some station zoologists have claimed, natural history museums were not unchangeable upholders of an outdated tradition. In fact, the natural history museum received a second wind in the

late nineteenth century, with numerous new foundations and a new conceptual dynamic. The period witnessed important museum reforms. These took
their inspiration from various sources, among which the new ideas about
nature propagated by the station movement. Some museums, indeed, would
set up their own stations, and take part in this movement themselves.

The transformation of the late-nineteenth-century natural history museum has been well studied. So far, however, historians have focused on the
changes in exhibition practices rather than research activities. They have
recounted how in the second half of the nineteenth century leading institutions, such as the British Museum in London or the National Museum
in Washington, "discovered" the masses and increasingly took up popular
exhibition as their central task. Research collections were separated from
those on public display, and, instead of endless showcases with taxonomically ordered specimens, a limited number of objects were selected to be
exhibited. And indeed several museum directors (particularly in the United
States) concentrated their activities on spectacular exhibits, but others
(particularly on the European Continent) continued to strive for scientific
prestige. Some of the latter hoped to revive the aura of their institution by
reorienting their work from taxonomy to the study of the animal in its natural environment. This is, for example, apparent in several German natural
history museums, like those in Berlin and Hamburg.[1] The clearest instance
of a museum that turned away from pedagogy in order to devote itself to
ecologically inspired science was located outside of Germany, however. This
museum is the Natural History Museum of Brussels.

The Brussels museum, the only of its kind in Belgium, drastically transformed its functioning in the period between the 1890s and the 1920s. It
was in these years that its director voiced the ambition to create a so-called
exploration museum—an ambition in which the creation of zoological stations took a central place. A metropolitan museum, so it was claimed in
Brussels, should incorporate exploratory field science. More even, it should
take it as its most important task. This chapter will research how the project
of the exploration museum was given form in Brussels, and which scientific
and museological practices it generated. In doing so, it will explain how a
metropolitan museum could be included in the station movement.

FROM TAXONOMY TO ECOLOGY

In 1914, the then director of the Brussels Natural History Museum, Gustave
Gilson, issued a lengthy book in which he elaborated on the mission of

his institution. The book was ambitious in design and elaborate in style. One crucial passage summed up its major message: "The Museum will have the heavy task to carry out the grand mission that constitutes its reason of existence! [. . .] It is an overwhelming duty, a gigantic and never ending work, of which nothing can distract it, and which will only be understood and appreciated once the mentality of the common public has changed, by yelling in its ears, one hundred times or more: Exploration! Exploration! Exploration!"[2] Gilson's vision can be summed up very briefly: the museum personnel had to leave the safe walls of its institution and set up expeditions in the surrounding lands. The museum, Gilson kept repeating, could not survive without fieldwork—could not survive without studying nature *in* stations situated in nature itself.

Gilson's call for reform was not new. It had gradually developed in a period of more than fifty years. It aimed at breaking away from the taxonomic cabinet that the museum of Brussels had traditionally been. The history of this cabinet had been long and is somewhat complex. Its oldest collections dated back to the eighteenth-century cabinet of the Austrian governor Charles Alexander of Lorraine; after it was transformed to a city museum, the collections were finally taken over by the Belgian state in 1846.[3] The ambitions of this state museum hardly differed from other mid-nineteenth-century natural history museums. The central aim was the collection, description, classification, labeling, and exhibition of objects. The model (both conceptually and architecturally) was the library—a place from which the museum echoed the aspirations of completeness and rational order.[4] The interest of the curators particularly concerned inventory: stratigraphy of geological layers, taxonomy, and comparative anatomy of living beings. Georges Cuvier was seen as the exemplary museum zoologist and his Muséum d'Histoire Naturelle in Paris as the scientific point of reference.[5]

From the annual reports of the Brussels museum in the midcentury it stands out clearly that the main point of attention was collection formation. The collection interests were very broad, both in terms of geography and subjects. Only gradually the focus was narrowed. Attempts in the 1870s to integrate the National Botanic Garden into the museum were unsuccessful, after which botany definitively fell outside the museum's scope.[6] Geology was for a long time one of the focal points, but in 1886 the Service of the Geological Map of Belgium was reorganized and withdrawn from the museum.[7] Out of necessity, the remaining disciplines (zoology, paleontology, and prehistoric archaeology) became the museum's spearheads in the following decades.

Because of the long-time focus on collecting, the personnel of the Brussels museum in midcentury consisted mostly of typical cabinet savants—exactly of the type that Gilson would later execrate in no uncertain terms. It was not so much by exploration, but by purchase and legacies of private collectors that naturalist objects were gathered. If the museum's personnel engaged in excursions at all, it was in their leisure time.[8] In Brussels one followed Cuvier's adage, which stated that "it is really only in one's study that one can freely roam throughout the universe." [9]

The first initiatives for reform were launched during the long directorship of the (ideologically volatile, but scientifically determined) geologist Edouard Dupont. From the moment he was appointed director of the museum in 1868, Dupont narrowed its geographical focus from the entire globe to Belgium's national territory.[10] Additionally, he tried to set exploration, particularly in the fields of archeology and geology, on the agenda. Dupont, however, also invested a lot of time in the discussions over the reissue of the Geological Map of Belgium, the practical worries that came with the move of his museum to a new part of the city, and the escalating quarrels with his scientific personnel.[11] All this would postpone the scientific reorientation he had envisioned. Only by the 1890s, he gradually started to realize his earlier ideals, and entomological, marine, and limnological explorations were subsequently set up.[12] Dupont envisioned a research museum that would be complementary to the country's university laboratories. For that reason he explicitly stressed his institution would not invest in academically established disciplines such as comparative anatomy, histology, and embryology.[13] It was not the laboratory but the field that according to Dupont was to be the workplace of the modern museum curator.

In annual reports and correspondence the museum's outdoor excursions were originally indicated simply as "explorations," but in the early twentieth century the more specific term *explorations éthologiques* [or: ecological explorations] came in use.[14] This terminology betrays its inspiration by the work of Alfred Giard and his circle at the marine station in Wimereux. The first to pick up the ecological program in Brussels, the entomologist Guillaume Severin, had indeed been in contact with Giard since the 1880s and befriended several Belgian Giardists. Severin was not trained as a zoologist, but as a draftsman, and he was probably the last practical naturalist to be hired as a curator at the museum. He was one of its only staff members to maintain good relations with Dupont, which was important for the fate of ecological work at the institution. Dupont openly supported Severin's approach and tried to extend it toward other disciplines.[15] In this, he did not so much call upon his permanent personnel (with whom he was often at odds), but rather engaged so-called national researchers—scientists from outside of the museum, who performed temporary exploration assignments.[16] This unusual strategy was meant to force a drastic break in the museum's research traditions.

One of the national researchers to be hired was Ernest Rousseau, an exuberant physician from a wealthy background and a friend of Severin's. His father, Ernest Rousseau senior, was a former rector of the University of Brussels, and his house was a well-known salon for the capital's left-wing intellectual and artistic elite.[17] By hiring the son, Dupont hoped to mobilize the network (and eventually the capital) of the father. Ernest junior specialized in ecological aspects of subsequently entomology, marine zoology, and limnology, subjects for which he was able to trigger a lot of amateur interest. The work of these amateurs was also put to use by the museum.[18]

More than Rousseau, however, Gustave Gilson would become the figurehead of exploration science at the Brussels museum. He was the most weighty national scientist to be attracted by Dupont, and he would eventually succeed the latter as director. A zoology professor at the Catholic University of Louvain, Gilson belonged to the political right—a faction in which the pragmatic Dupont had good friends.[19] Yet scientifically the choice was an awkward one. Dupont contacted Gilson with the request to set up an ecological study of the Belgian marine fauna, but, at that point, the latter had little if any experience in fieldwork. Specialized in microscopic histology, Gilson was known as a typical laboratory scientist. His biotope was one of microtomes and modern staining liquids, and a colleague described his early work as "cutting up everything that presents itself."[20] Yet Dupont's request led to a conversion. From the late 1890s onward, Gilson devoted himself entirely to marine zoological and oceanographic expeditions, and became one of the most important Belgian defenders of the importance of fieldwork. His ecological interest brought him into contact with the Wimereux circle, and Giard's *Bulletin Scientifique* would eventually issue a long article by his hand on a parasitic wood louse. Such a publication was a rare honor for a Catholic.[21]

By the beginning of the twentieth century, Dupont had thus created a small team that had to give the museum's zoology department a new ecological dynamic. It consisted of a draftsman-turned-entomologist, a physician-turned-limnologist, and a histologist-turned-oceanographer. What these men shared, despite their differing backgrounds, was a strong belief that organisms could be understood only by studying them in their natural habitat.

This conviction turned their eye to foreign zoological stations (including those of Naples, Wimereux, and Plön), and quickly they would link up the museum with the station movement.

URBAN BOURGEOISIE AND FIELD BIOLOGY

The ecological exploration program of the Brussels museum was set up in the aftermath of its move in 1892. From the center of Brussels, the institution was relocated to a new building in the Leopold Park, situated in a residential eastern part of the city. The move had symbolical meaning. It followed the earlier exodus of the Brussels aristocracy and *grande bourgeoisie* in the same direction. The museum would be located in the center of the quartier Leopold, a checkerboard of mansions conceived in the late 1830s by a coalition of royalty and heavyweights of Brussels financing.[22] Their ideals were reflected in the street names, which next to a *rue de l'industrie* and a *rue de la commerce* also included a *rue de la science*. The area had a particular social fabric, which could provide the museum with its self-proclaimed target audience "of an intellectual superior culture."[23] It was furthermore close to the heart of Belgian politics, on which the institution depended financially. In the immediate vicinity of the museum one also found the Academy, the seats of most important naturalist societies, and the University of Brussels—the latter being an important early center of field biological studies. Furthermore, most Brussels scientists and intellectuals had their houses in the direct vicinity of the Natural History Museum.[24] Ernest Rousseau senior, for example, held his salon just across the street.

The cityscape of the quartier Leopold obviously offered few prospects to perform ecological field research. Yet it was in this urban milieu that "unspoiled" nature was intellectually constructed as a place of value and scientific interest. From the late-nineteenth-century urban perspective, nature's wildness seemed increasingly vulnerable, particularly in a country that was one of the most industrialized and urbanized in the world. The Natural History Museum itself would testify about Belgium's vanishing wildness by exhibiting, among others, two of the country's last wolves, shot by King Leopold I in the 1840s.[25] By the turn of the century, the urban bourgeoisie showed an increasing interest in what was still left. Nature excursions and hiking became increasingly popular, preservationist ideas started to spread, and calls were heard to scientifically map and study the remaining "original" landscapes of the country. Many of these initiatives revolved around a small group of scientists and intellectuals who resided in and around the

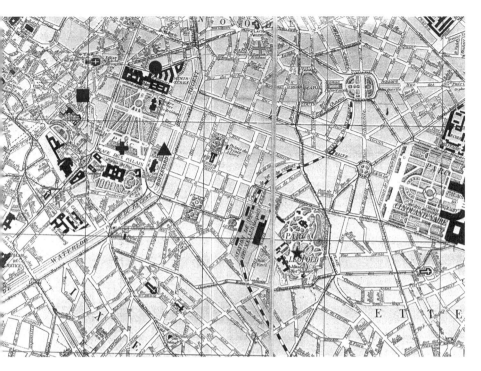

FIGURE 7.1 The natural history museum in the Parc Léopold (star) was connected to the rest of the country via the close-by railway station. Via the Quartier Léopold it was linked up with some of the most important political and scientific centers in Belgium: the Parliament (circle), the Royal Palace (cross), the Academy (triangle), and the University of Brussels (square). From *Plan générale de Bruxelles*, 1908.

Quartier Leopold.[26] It was also the patches of nature to which they brought attention that would constitute the work domain of the museum's ecological program.

The exploration museum was not designed to perform ecological research in situ. Rather it was a hub from which to reach the rest of the country. From the Quartier Leopold railway station, Belgium's dense railroad system could bring curators quickly to even the most remote areas of the national territory. At first, this possibility was used only sparsely, but by the turn of the century it had become a current practice. Severin's day excursions, for example, increased from just four in 1892 to ninety-nine in 1900.[27] Yet, like we have seen elsewhere, such an itinerant study of the interactions between organisms and their milieu, had several disadvantages. Numerous aspects of the organism's life could, after all, be studied only by prolonged and repeated visits. Furthermore, much of the scientific equipment needed could be transported only with "tiresome efforts by the staff."[28] Again, spatial strategies were designed to overcome these problems.

One strategy consisted in moving parts of nature to the metropolis, rather than the other way around. Severin, for instance, had whole insect nests, cadavers, and tree trunks transported to the museum building. The "ensembles" were then studied in makeshift terrariums during the following weeks, among others to trace morphological variation within insect species.[29] In order to provide the isolated objects with a broader context, elements of the surrounding landscape were turned as much as possible into "immutable mobiles" and brought to the museum as well. Drawings were made, casts were produced, and pictures taken. Particularly field photography would be taken up for its presumed "absolute absence of imagination."[30] Yet the possibilities this combination of strategies offered for the study of living nature were limited, particularly since the government was not willing to invest in an infrastructure for keeping live animals at the museum. In 1923, Gilson still complained to the Minister that the museum did not possess its own permanent terrariums and aquariums. Hopes to set up a small glasshouse and a garden for the breeding of insects had proved equally vain.[31]

Rather than building out a new infrastructure itself, the museum eventually resorted to collaboration with private organizations. One of these was the majestic Public Aquarium of Brussels. The aquarium, which opened its doors in 1906 on the fancy Avenue Louise, was set up by the fish-breeder Charley Poutiau and the twin brothers and barons Constant and Auguste Goffinet. The Goffinet brothers were close aides of King Leopold II as well as owners of a large estate, where they practiced "methodical fish culture." Poutiau for his part led a company self-importantly called Establishment for Aquaculture, Fish Culture, Scientific and Sportive Hydrobiology that covered some 1000 hectares of water. The goal of their common Brussels aquarium, then, was to popularize knowledge on the Belgian freshwater world and promote repopulation projects for Belgian rivers. As such, the aquarium can be seen as a late echo of the imperial aquaculture projects in mid-nineteenth-century France. The Natural History Museum closely associated itself with the undertaking. Gilson and Rousseau acted as advisors to the aquarium, at which they were allowed to use the laboratories and reservoirs. This gave them the opportunity to observe the behavior of aquatic animals and set up the breeding of certain species—which was impossible in the museum itself. The aquarium made nature accessible in the urban space, and this for both the bourgeois visitor and the scientist.[32] For both, it constituted a place of retreat. Visitors noted the silence in the building and the atmospheric "greenish light that shone through the aquarium plants."[33]

Despite the practical possibilities it offered, the Brussels aquarium (not unlike the Naples aquarium discussed previously) remained a re-creation of nature rather than the real thing. An exploration museum that proclaimed one had to observe organisms "above all in the natural conditions of their environment" needed other instruments.[34] What it needed was biological stations. Because of the scarcity of finances, the museum's director again could establish these only through cooperation with so-called private installations. The first step in this direction was taken in the 1890s. In these years, two professors of the University of Brussels, the botanist (and Giardist) Jean Massart and his zoologist colleague Auguste Lameere, started to set up a yearly "ambulant laboratory" for their students. Each summer, they traveled to a remote area in the province with expedition boxes full of microscopes, reagents, and books. Although organized by the university, also museum people like Severin participated in this combination of a broadly conceived exploration, a rite of passage, and a summer camp. Severin's photo archive furthermore shows that he also made use of other improvised field laboratories set up by colleague scientists.[35] In this way, the museum could participate in more prolonged exploration work without heavily investing in it.

Preferably, these private, provisional labs were put up in landscapes that could serve as natural laboratories. The Scheldt estuary, for instance, was a popular destination because it offered brackish water of a decreasing salinity. The landscape thus offered a rather straightforward variable, of which the impact on the spread, behavior, and adaptations of the local fauna could be studied in natural conditions.[36] Such a strategy (and indeed the ecology that came with it) was very similar to that of the French Giardists. But also other strategies were used. The High Fens, for example, with its relatively cold climate offered an isolated area of subalpine fauna and flora, which could be researched as a rather well-delineated group of interacting plants and animals.[37] Like the lake it offered a microcosm, easily studied as a life community in the German tradition. Hence, the actual approach that was deployed depended on the landscape, but in both cases ambulant laboratories were believed to be necessary.

Next to ambulant laboratories, permanent biological stations were set up. In the words of Gilson these had to be "established in well-chosen localities, where one expects the exploration has to be perpetuated."[38] Eventually, the Brussels museum would become engaged in the functioning of two of such stations: the marine laboratory in Ostend and the freshwater station in Overmeire. Both could be realized only thanks to improvising. Around 1900, Gilson set up a temporary laboratory in the Free Fishery School of Ostend, while collaborating with the public aquarium in the same town for the inventory of the coastal fauna.[39] On the eve of World War II, then, he temporarily had his laboratory in an oyster farm, while for the study of the full sea he could make use of ships that were put at his disposal by the Ministry of Interior and Marine.[40] The whole undertaking continued to carry a provisory character, and all in all Gilson's infrastructure compared badly with that of foreign marine stations. Yet even his modest marine zoological work was relatively expensive. In 1907, for instance, his research accounted for more than 60 percent of the museum's expenses.[41] The fact that similar expenditures endured in the following years, indicates that the then leaders of the museum saw them as crucial in the modernization of their institution.

The freshwater research of the museum, then, was strongly stimulated by the foundation of the station in Overmeire in 1906. This foundation was a private initiative of Rousseau. Thanks to investment of a Maecenas, he was able to set up a laboratory in a former hotel, situated in a marshy region. The place for the station (which mimicked the station in Plön) was chosen for its variety in species and its easy accessibility by rail from the most important scientific centers of the country. Next to a laboratory, the research station had a small library and a few aquariums. It was open to all researchers, but its specimens were systematically sent to Brussels, where they would constitute the core of the museum's ecological collection of freshwater biology. Like many other stations, Overmeire had its own journal, the internationally oriented *Annales de Biologie Lacustre*. Among its contributors were leaders of the discipline such as François-Alphonse Forel and August Thienemann, and the journal quickly counted as trendsetting. Through Rousseau the whole undertaking remained strongly connected to the museum.[42] Just as in the case of the Avenue Louise Aquarium, the ambulant laboratories, and the Ostend Marine Station, costs for the museum remained limited thanks to an improvised alliance with private initiatives.

Studying nature in nature, the visitors of research stations quickly discovered, was particularly problematic in a country such as Belgium. As one of the guests of the ambulant laboratory noted, "wild land" was becoming increasingly scarce at the turn of the century because of the pressure of industry and agriculture. He added that this wild land was to be studied before it disappeared altogether.[43] The changes of Belgium's natural landscape

A)

B)

D)

C)

FIGURE 7.2 Annexes of the Brussels museum: A, the Avenue Louise aquarium, B, the ambulant laboratory of the University of Brussels, C, the marine station in Ostend, and D, the
limnological station in Overmeire. From Gilson, *Le musée* (1914), 208; Archives de Université
Libre de Bruxelles; Gilson, *Le musée* (1914), 198; Author's personal collection, respectively.

and the consequences of these changes would eventually become a subject
matter in their own right. Severin, for instance, studied the insect plagues
related to the increasingly uniform monocultures of southern Belgium's forestry.[44] Rousseau worked on the ever more visible effects of the pollution
of rivers, and devoted a permanent exhibit to the subject in the Brussels
aquarium.[45] Gilson, then, showed a sustained interest for the decline of fish
stocks in the North Sea and how this related to the mesh size of the nets
used by Belgian fishermen.[46] Zoological stations might have been meant
as places of retreat in unspoiled nature, but this did certainly not exclude
visiting zoologists from studying the increasing influence of human activity.

LOCAL, PURE, AND HEROIC SCIENCE

The human interference with nature constituted only one of the many subjects the annexes of the Brussels museum were set up to tackle. Most of the research continued to concern nature in its (perceivedly) unaffected form. There was attention to the life cycles and metamorphoses of water insects, the nesting behavior of birds, the seasonal dimorphism and the migration of marine species, and the feeding habits and the natural enemies of freshwater fish.[47] These interests were all rather new in the Belgian context. As such, they necessitated a new type of researcher and new scientific approaches.

A good example is offered by Gilson's studies of the migration of commercially exploitable fish in relation to their milieu. To monitor the latter, Gilson's oceanographic expeditions systematically measured the salinity and temperature of the sea water. Inspired by work performed at the Plymouth Marine Laboratory, Gilson furthermore set up a project to study sea currents, which included releasing small floats in the hope that finders would send them from far-off beaches back to Brussels.[48] The strategy, which depended on the participation of nonscientists in the field, obviously reminds of the contemporaneous (but independently developed) bird-banding projects. This is even more the case for Gilson's attempts to track the movement of the fish themselves by marking them with tags.[49] This procedure, introduced by the Dane Johannes Petersen in 1899, quickly became popular among European marine zoologists, thus replacing the older biometric techniques to identify migrating herrings developed at the Helgoland station. The foundation of the International Commission for the Exploration of the Seas (ICES) in 1902, which Belgium entered the year after, largely stimulated this kind of work.[50] Because Gilson soon was appointed the Belgian spokesperson to the ICES, his museum and its marine station became the national centers for carrying out the Commission's research program. This role obviously matched well with the aura of exploration the museum was seeking for itself.

The collaboration with the ICES nicely illustrates how the Brussels museum tried to combine an international orientation with strong local roots. Ecological observations required a good field knowledge, which, according to the Brussels scientists, could be acquired only when the actual geographical scope of the research ambitions remained limited. Unlike the highly publicized long-distance expeditions that were carried out by vessels such as the Challenger and the National, Gilson's oceanography was limited to only a small zone along the Belgian coast. Also the museum's inland explorations remained restricted to a few areas, which were thought to be representative for the national territory. This, Gilson insisted, had nothing to do with "narrow national particularism," but everything to do with the ambition to understand organisms in their actual habitat. The universal laws of field biology, so he claimed, could be produced only by local expertise.[51] With such a claim, Gilson explicitly distanced himself from those who wanted to root claims of universality in placeless laboratories. Historians and sociologists of science have described how this rhetoric of placelessness, of eliminating all influence of locality, developed into the hallmark of modern universalistic science. In this rhetoric the field represents only the *here*, while the laboratory offers a generalized *anywhere*. Gilson (and the station movement at large) countered this discourse.[52] The field was not a place of mere anecdote, he claimed, but a site that, if approached correctly and by the right people, could give access to universal truths.

This conviction impacted on the human resource management of the museum. It implied that fieldwork had become a high-status activity, and was to be entrusted to the higher-ranking museum staff (including the director himself). Of course, the museum always had engaged fieldworkers, but their scientific prestige had for a long time been very limited. It concerned hunters and collectors, who mostly acted outside of the framework and the control of the museum, and their collected specimens reached the museum only thanks to the intermediary of naturalist-merchants.[53] The information on the wider context in which these objects were gathered remained very limited and was seen as untrustworthy. But when the ecological approach highly increased the significance of this wider context, this situation was obviously no longer tenable. Gilson therefore stressed that fieldwork was not to be handed over to "mediocre travelers or undefined amateurs." Free nature confronted the researcher with so many "pitfalls and unexpected events" that only *real* scientists could make precise (and universally valid) observations there.[54] As early as 1907, the museum invested ten times more in explorations by its staff than it did in the direct purchase of specimens.[55] This reorganization of research at the Brussels museum does not seem to have been an isolated phenomenon—nor was it limited to zoology, for that matter. Andrew Zimmerman has described how, in exactly the same period, German anthropologists exchanged research in museum cabinets for stationary fieldwork. Also here one of the main arguments was that the information delivered by undefined and hasty travelers was not to be trusted.[56]

While the reorganization of the Brussels Natural History Museum can be read as a history of professionalization, it was also clear to its initiators that the ecological project could not succeed with professional scientists alone. It was necessary, so it was stressed in the museum's annual reports, to be in touch with local volunteers such as landowners, hunters, or wanderers on the beach. Ecological observations often depended on circumstances that only rarely occurred, so that it was essential to have eyes and ears on the scene. Curators prided themselves with knowing these local informants personally. Contrary to the nineteenth-century travelers and amateurs who provided the museum with objects, these informants had to be checked for their trustworthiness—from which the success of the ecological project depended.[57] The new Brussels museum thus relied on an economy of trust very similar to that propagated by station zoologists such as Giard, Zacharias, and Thienemann.

The coalition of professional metropolitan scientists and local amateur informants was, according to the museum's directors, engaged in a project of "pure" science. Throughout his directorship, Gilson would stress that economic applications of naturalist research (or: "the art to apply Science to the exploitation of Nature") did not belong to the museum's responsibilities.[58] Like other proponents of the station movement, however, the Brussels researchers (including Gilson) combined this explicit choice for fundamental ecological science with regular references to the economic prospects of their work.[59] At several occasions, the museum's personnel also happily put its ecological expertise at the disposal of commissions and ministries whose aims were chiefly economic. Gilson's active engagement in the attempts of the ICES to manage the European fisheries is a case in point. Rousseau, for his part, was hired to write reports for the Commission of Fish Culture on topics such as salmon breeding.[60] Severin, finally, performed entomological work for the Ministry of Agriculture and the Commission for Waters and Forests. In this context he researched, for example, the ecological causes of the invasion of the great spruce bark beetle in the Belgian pine forests.[61] Later, in response to a demand of the International Ornithological Congress in Paris, he was appointed to analyze more than 2000 bird stomachs (sent in by state foresters) to assess the birds' importance in the combat against insect pests, in this way continuing the early work Johannes Thienemann had performed in Rossitten.[62] Such developments in the direction of applied work meant that ecological explorations gradually gained an air of economic significance—albeit somewhat ambiguously combined with an official allegiance to pure science.[63]

It is not only through its possible economic advantages that ecological explorations gave the museum a new aura. More important even was its connotation of adventure, which became an increasingly important element in the self-presentation of the museum. Gilson played a key role in this image building, by stressing that his museum collaborators were not "savants," but rather "men of action." As a director, Gilson (like indeed many station zoologists) wanted to get rid of the cliché in which naturalists were represented as maniacs interested only in reconstructing skeletons and pinning insects. For the same reason he wanted to change the official name of his museum's senior staff. Traditionally Brussels curators were called "conservator," a term Gilson wanted to replace with "exploration leader." The linguistic shift had to indicate that the museum's researchers were complete scientists—"capable to perform intellectual work amidst the fatigue, the deprivation, the dangers and the fights which are inextricably linked with research on land and sea, in laboratories, on ships, in woods, marshes, fields and mountains." The term "exploration leader," Gilson added, "would exclude those men from the higher staff of the museum, who never took part in the rough campaigns on the warpath of Man's conquests on Nature."[64] The rhetoric stuck. At a celebration in his honor in 1936 Gilson's successor at the museum, Victor van Straelen, stressed that the actual workplace of Gilson never had been the auditorium, but rather the open sea and the jungle. The latter was nothing but an extrapolation on the basis of Gilson's adventurous self-image. Except for a one-time expedition to the Fiji islands, he had, after all, never left Europe.[65]

Exploration implied suffering, so it was heard at the Brussels museum. In one of his reports to the Minister of Interior, Dupont stated that expeditions always entailed a "tough life." He referred among others to the wintry explorations in the Ardennes in the southern part of the country, during which the museum's curators at the ambulant laboratory were systematically exposed to freezing temperatures.[66] Gilson later expanded on the image of the suffering scientist. In 1914 he wrote: "Science is without doubt a stepmother. She does not promise happiness to Humanity, and, from her followers, she asks that they forget the notion itself."[67] And he described his own research carried out at the marine station as "slow, hard and painful [work] that often takes the character of a fight against the elements and that is often surrounded by danger."[68] Scientists, he believed, should be undaunted, and it was also in that way that Gilson would be remembered himself. In the above-cited speech, Van Straelen invoked the image of Gilson as a sea dog "shaken on a small ship, devoid of all comfort, [. . .] indifferent to all danger."[69] The

FIGURE 7.3 Gustave Gilson, posing as a manly "exploration leader." (Courtesy of Leuven Universiteitsarchief, Topografisch-Historische Atlas)

description echoes a photograph Gilson had posed for. It shows him in seaman's outfit, staring in the distance, his arms manfully crossed.[70]

The image of the suffering scientist who sacrificed his own happiness and even his physical health while performing his work was obviously not a new creation of Dupont or Gilson. In the late nineteenth century the idea of scientific martyrdom had become increasingly popular throughout the whole Western world, and in this the exploration discourse had been particularly instrumental. More and more activities (even of chemists or historians) were described as constituting a form of "exploration." Whereas the naturalists of earlier centuries had often distanced themselves from the dirty drudgery and physical discomforts that came with explorations, these elements now became key to the scientific self-image.[71] Station zoologists made sure to pick up on this discourse—albeit to different degrees. Giard and Zacharias used the idea of the discomforted scientist relatively sparsely, while it was very explicit in the writings of Johannes Thienemann, and even more so in the work of Gilson. In reality, the discomforts for the self-proclaimed Brussels "exploration leaders" seem to have remained rather limited. The many exploration pictures of Severin, for example, bring to mind a picnic rather than a life-threatening expedition.

EXHIBITION FORMS AND AMATEUR PRACTICES

The "manly" ecological research, developed in ambulant and permanent field stations, influenced the exhibition practices in the museum itself. The endless galleries of specimens that had been designed in the nineteenth century no longer represented the research practices of the museum and its stations around 1900. The ambulant laboratories, the limnological and the marine station, and the urban aquarium provided information on the interaction of organisms with their environment, which could be communicated only by alternative modes of representation. In this, the Brussels museum could take inspiration from several foreign examples. By the end of the nineteenth century numerous natural history museums were experimenting with new forms of exposition that were purportedly more didactic and scientific. As indicated there was a growing consensus that museums no longer had to display their complete collection, but rather a small selection from it (banishing the bulk of their objects to storerooms). Moreover, curators sought to present naturalist specimens in more attractive ways. In zoology departments the diorama caught on. Dioramas showed the stuffed animal in a recreated landscape, often with a painted backdrop.[72] At first sight, such a representational mode seems to perfectly meet the aspirations of the ecologist. After all it reconstructs the natural context in which the represented animals had lived. For several reasons, however, the Brussels curators were far from enthusiastic about this new exhibition form.

Severin had been the first Brussels curator to come into contact with the diorama during his study travels to museums in Central Europe and the United States (in 1898 and 1906, respectively).[73] His impression was highly negative. Dioramas were expensive and unscientific, so he claimed. They offered "a more or less colored, so-called esthetic representation" that was rather suited to trigger a feeling of sensation than to convey scientific information.[74] This feeling matched with an outspoken aversion on the part of Gilson toward the "didactism" that was purportedly typical of American natural history museums.[75] This aversion was not exceptional among European

curators. Particularly museum curators who still carried scientific ambitions tried to avoid exhibition forms that were considered too populist.[76]
In a period in which university laboratories already seemed to outshine the
scientific work at museums, curators wanted to avoid by all means that a too
"sensationalist" museology would lead to further loss of scientific prestige.[77]

The Brussels directorship was nonetheless aware that new exhibition
practices were necessary. The research in ambulant and permanent stations
had brought new questions to the fore, and had brought a new kind of material to the museum. Rather than using dioramas, the directorship hoped to
exhibit this material through sober ecological displays. The target audience
they had in mind for this was "the literate visitor" with a "higher intellectual
culture," but not specialized in the sciences. This literate visitor was to be
initiated in the morphology of the exhibited species *and* their life conditions.[78] The first series of ecological collections to be actually realized were
the entomological series composed by Severin in the late 1890s.[79] He designed what he described as "ecological boxes" that had to stimulate an interest in animal ecology. These boxes showed, for example, butterfly species
in several variants, their caterpillars, and the vegetation on which they were
found. Placards further explained the shown morphological and ecological
themes, and this with detail that was unusual in contemporary museums.[80]
Furthermore insect nests (such as ant colonies) were exhibited. In order to
get the fragile constructions to the museum a private firm was even hired to
freeze the nests in the field before transporting them to Brussels.[81]

Although Severin and Dupont eagerly stressed the innovative aspects of
their ecological museology, it was clear that they had found their inspiration
at least partially elsewhere in Europe. Severin's travel in 1898 to several
Central and Southern European countries was crucial in this respect, since
it confronted him with a whole series of attempts to translate holistic visions of nature in museum exhibits. On his trip, Severin must have seen
among others the oyster bank and coral reef exhibits that Karl Möbius had
set up in the Museum für Naturkunde in Berlin as illustrations of an ecological *Lebensgemeinschaft*. Yet these new exhibition forms seem not to have
impressed him, since they do not figure among the numerous photographs
he made on his trip.[82] The Brussels ecology tradition, after all, was in line
with Giard's and was interested in the individual's interaction with the environment rather than the study of life communities initiated by Möbius.
If the ecological boxes took inspiration from Berlin, it was rather from the
exhibits in economic natural history in which, among others, harmful insects were paired with the plants they lived on. Severin, furthermore, saw

similar exhibition forms in several novel museums for economic entomology and forestry.[83] It was with these applied fields of research, rather than the more theoretical German ecology, that the Brussels exploration museum would identify.

The first ecological boxes were displayed at the Brussels museum at the end of Dupont's directorship. They were accommodated in the new wing of the museum, which was opened in 1905 and which impressed with its Art Nouveau style. The prestigious building was used to exhibit the Belgian collections of the museum: the vertebrates on the ground floor, the invertebrates on the first floor. The design of the building antedated the "ecological" orientation of the museum's research and was, like many other museum buildings, inspired by geological chronology. A system of stairs throughout the building indicated geological time. The entrance was in the hall of the Quaternary, in which (rather revolutionarily) also prehistoric man had been integrated.[84] Three steps up, in the same space, the visitor found himself in the Tertiary, again a few steps higher, fossils from the Mesozoic were exhibited.[85] The progressive, evolutionary story that structured many nineteenth-century natural history museums was thus told backward in Brussels.[86] The visitor returned in geological time, experiencing an evolution in reverse.[87] This story was not fundamentally altered by the introduction of ecological boxes, but they did establish a new scientific sensibility. Rather than showing an evolution through geological time, they illustrated an evolutionary adaptation to a specific milieu on one precise moment.[88]

After the first ecological boxes on insects had been established, others with regard to marine and freshwater fauna were added. As such they clearly reflected the research that had been performed in the stations annexed to the museum.[89] The Minister of Interior encouraged the museum's director to extend this approach to mammals and birds, shown in their natural habitat. Dupont, however, tried to ward off the suggestion. He argued that it was very difficult to photograph these animals in action and to collect enough ecological information to make such groups scientifically sound. Ecological boxes were impossible to design for these groups, and sensational dioramas were never considered an option.[90] Severin supported this line of thinking. In a memorandum he stressed that preparatory studies to design mammal habitat groups were extremely expensive and thus within the reach only of wealthy institutions such as the American Museum of Natural History. In Brussels they could be performed only at the expense of the budgets for exploration, which lacked finances already. According to Severin, people who wanted to see mammals in action better went to the zoo.[91]

The new museology might not have been aiming at the masses, but still it was definitely broadening the museum's audience. In similar ways, the scientific work of the museum became more and more inclusive in the years around 1900. The groups of nonscientists who engaged in the work of the museum increased. Next to hunters and landowners, the ecologists were in touch with oyster-breeders, rangers, and brigadiers.[92] Ambulant stations furthermore could not be erected without contacts with local notables or other authority figures.[93] Finally, most curators were active in scientific societies, which for decades constituted the most crucial point of contact between amateurs and professional scientists.[94] It was the same amateurs the museum hoped to engage in the labor-intensive projects that the ecological approach demanded.[95] At the same time, the museum's staff explicitly strove after a "professional" image for itself. To the outer world they stressed that their methods strongly differed from the somewhat random collecting purportedly still in vogue in amateur circles. The ecological approach might have increased the contacts with amateurs, but at the same time they were treated with an increasingly paternalistic attitude.[96]

In the same context, one can situate the rise of the popularizing scientific excursion in Belgium. Its instigator was (as discussed in chapter 4) Jean Massart, who, in the early 1910s, set up a series of popular nature hikes. Also in the intellectual circles of the Natural History Museum similar excursions would soon be organized. In the aftermath of the First World War a central role in the organization of these would be played by the newly founded society *Les naturalistes belges* [The Belgian naturalists]. Rousseau was cofounder and one of the driving forces of this association, which had grown from the National Society of Aquarium and Terrarium Amateurs. The latter society was founded in 1916, but quickly transformed from a small group of aquarium lovers into a general society for the vulgarization of science. The society organized exhibitions, screenings of documentaries, lectures, and excursions. The enterprise was a success. In 1921 the society counted 2500 members, which made it the biggest naturalist society of the country.[97]

The inspiration of *Les naturalistes belges* was, not surprisingly, ecological. Its founders wanted to transfer the ongoing reform from the Natural History Museum in Brussels to the circuit of nature enthusiasts. One hoped to reorient the activities of the latter from the collection of dead species to the observation of organisms in their habitat. Or with the words of the society's president, the Brussels physician Léon DeKeyser: "We would want that the young naturalist no longer satisfies himself with catching an animal or with prudently picking a plant, to then identify it and put the one behind glass, and glue the other on a paper with a label in calligraphy. That is dead science. What we would like him to do is to observe an animal in its environment, with its habits and customs. That would be living science—so much more attractive, so much more compelling."[98] Hence, after having modernized the scientific work in the Natural History Museum, the ecologists also oriented the amateurs to their project. Visits to the field annexes of the museum were one of the recurrent activities that had to inculcate this ecological attitude.

THE LEGACY OF MUSEUM ECOLOGY

Around 1920, the new ecological look had been spread from a small scientific coterie, to a rather substantial audience. Yet in the same period, the success faltered. The reform of the Natural History Museum got stuck halfway. There may have been ecological collections on insects and freshwater biology on display, but the rest of the collection (unconnected to field stations) was largely left untouched.[99] Furthermore the scientific infrastructure of the museum had suffered heavily during the war and several well-trained staff members had fallen in battle.[100] For several years, there was little financial room to move.[101]

In 1925, Gilson was forced into retirement because he had reached the age of sixty-five. This was clearly against his will. Stubbornly he tried to postpone the decision with the Minister of Arts and Sciences. His project was not finished, he insisted. He argued that, if the reorganization of the museum ended halfway, his successor would have to start without a proper base, and yearlong efforts would bear no fruit.[102] Yet the administration stuck to its decision. Gilson was replaced without delay by Victor van Straelen—a Brussels geologist without ties to the museum. Embittered (and endowed with a small pension) Gilson left his office. He continued to dedicate himself to marine exploration, succeeded in receiving subsidies for a new laboratory in Ostend, and continued to work for the ICES. The Marine Institute of Ostend remained under his leadership until he died in 1944. The relations of this new institution with the Brussels museum were stern at best.[103]

Next to Gilson's retirement other factors played a role in the withering of the ecological program in Brussels. Rousseau—the key figure of ecological freshwater zoology—unexpectedly died in 1920. The laboratory of Overmeire came under the leadership of the Brussels zoologist Auguste Lameere, but was eventually shut down in 1926 when the most important individual sponsor of the station also died.[104] Severin retired in 1927 and nobody in the museum took over his ecological interest in insects. The aquarium in the Avenue Louise, then, remained closed for many years after the war. Thanks to state subsidies it could reopen in 1925, but in the 1930s the state of the building deteriorated and in 1937 it was definitively shut down.[105] Without functioning annexes in nature, the reform program was hard to carry through.

In this way, the exploration museum planned by Dupont and Gilson was never fully realized. Moreover, the national orientation of its research was eventually abandoned. In 1936, Van Straelen had to admit in one of his annual reports that the zoological exploration of the national territory had not profited from the general growth of the museum since his appointment.[106] A lot of attention did go to expeditions in the National Parks of Belgian Congo, of which Van Straelen was also the governor, but in this way the geographical coherence of the collections was lost. Furthermore, Van Straelen broke away from Gilson's aversion of didactics. During his directorship an educative service was set up, and eventually even the dioramas that once were anathema made their entry to the museum.[107]

All this does not mean that Gilson's ecological program in Brussels had been a failure. In Belgium, it was after all the small group of Brussels museum curators that, in the early 1900s, formed the core of the station movement. This group not only constituted the heart of the Belgian professional scientists who took up the study of animals in their natural habitat, but also spread their vision to a broader audience through their museum exhibits and their participation in the Brussels aquarium. With the creation of *Les naturalistes belges,* they stimulated this audience to actually perform ecologically inspired research. Internationally the exploration museum might have found little imitation, but this does not mean its work went unnoticed or was carried out in isolation. Gilson's mission statement met with acclaim abroad, among others in *Nature* and *La revue scientifique.*[108] Rousseau's *Annales de biologie lacustre* was internationally referenced and it played a significant role in creating an international research community of limnologists in a crucial phase of the discipline. Severin's fieldwork, then, was in line with developments of economic entomology in the United States, where he was praised as "a man of great acquirements."[109] Maybe Brussels was no leader of worldwide museum reform, but it was certainly recognized as a place of serious science. In the context of Belgian zoology it acted as an uncontested center, and within the national borders its intellectual orientation was highly influential.

The developments at the Natural History Museum of Brussels may have been to a certain degree idiosyncratic, but at the same time they were illustrative of a museum world that witnessed rapid reform, growth, and diversification. Unlike what outsiders may have claimed, museum zoology around 1900 was not a static and old-fashioned project, but a dynamic enterprise of growing importance and scope. Museum curators actively responded to the changing institutional constellation of the life sciences, and they did so with a varied set of strategies. While several American museums played the pedagogical card, many of their European counterparts hoped to regain scientific status. Of the Europeans, the curators of the Natural History Museum in Brussels were the most unequivocal in their refusal of the aura of pedagogy. Their strategy was to invest in serious science that would be complementary to that of the university laboratory. This strategy involved sending curators to the field, deploying stations as their workplace, and developing new exhibition practices to display the scientific results the stations procured.

In a period in which museum work witnessed a loss of epistemic authority, the foundation of field stations seemed to offer at least a partial solution. For this reason, it is not so surprising that natural history museums became integrated in the station movement. In some cases of newly founded institutions—such as the Oceanographical Museum in Monaco or the North Sea Museum in Helgoland, both discussed in chapter 1—the museum and the station were closely integrated from the start. The example of Brussels, however, shows that the station movement also managed to transform existing institutions. The initiative of setting up annexes in nature (whether ambulant or permanent) altered the research that was performed at the museum, the expositions it housed, and the self-image of the people who worked there. In this way a metropolitan museum could play its part in promoting the study of nature *in* nature.

Residents in the Field

The historian of science David N. Livingstone has observed that, in the field, "the investigator is likely to be the visitor rather than the resident."[1] This observation might be true in general, but it neglects the ventures that have been set up to accommodate scientists in the natural habitat of the organisms they are interested in. For the life sciences, these ventures culminated in the advent of the biological station in the 1870s. Since then, many life scientists have made nature their home, thus becoming residents in the field. It is the impact of this spatial shift on the study of animals that has been examined in this book.

Biological stations were diverse in their scale, population, research practices, and organizational framework. In some instances, stations were used as sites to acquire placeless knowledge—from which the references to the locality of research were erased as much as possible. Several stations were, after all, established as material incarnations of the "new zoology" (discussed in chapter 1), which was centered around placeless morphological and embryological research. But because of their geographical location, field stations clearly also offered research possibilities of a different kind. These possibilities were quickly turned into practice, and many stations became important centers of place-based research on the interaction of animals with their environment. This book claims that the rise of such place-based work in the life sciences was in fact closely intertwined with the advent of the biological station and the new practices it enabled.

Several early biological stations focused on scientific questions that concern what present-day scientists describe as the "ecology of place."[2] In fact, they were highly important (but so far little acknowledged) catalysts of the study of such questions. This is not to say that, in late-nineteenth- and early-twentieth-century Europe, biological stations were places of ecological discipline-building. The proto-ecological practices encountered at zoological stations were varied and went under different names: *éthologie, Oekologie* and bionomy, *Planktonkunde*, hydrobiology, the study of the *Lebensweise der Tiere* and limnology, marine zoology, ornithology and *Biologie im engeren Sinne*. Yet, despite a variation in terminology and concrete subject matters, these projects largely agreed about which general questions were scientifically interesting, about what was (epistemically and socially) good scientific practice, and about which type of scientist could deliver this practice. The proto-ecology that was developed in zoological stations focused on living animals in relation to each other and their environment. Its practitioners believed this subject could be studied *experimentally*, but they conceived this term broadly so that it did not exclude, but rather prioritized observational fieldwork. Experimental proto-ecological fieldwork was associated with patience and retreat, with a practical mind and manly perseverance. Often it was construed as a project that was oppositional to trends in the "mainstream" and "elite" sciences, which it was ultimately bent on reforming.

Not in all biological stations did the ecological interest materialize into concrete research. In the iconic Stazione Zoologica of Anton Dohrn in Naples the success of laboratory-oriented morphology and embryology overshadowed the study of animal life habits. Much to Dohrn's own concern, placeless indoor work outcompeted place-based outdoor studies in late-nineteenth-century Naples. Yet other stations showed a different dynamic. The stations of Wimereux, Plön, Rossitten, Overmeire, and Ostend—all dealt with in this book—are illustrative in this regard. Place-based research was not only integrated in the stations' programs, but explicitly presented as their prime reason of existence.

This focus on *local* nature set the mentioned stations aside from several other venues of the life sciences. Most botanical gardens, zoos, natural history museums, and laboratories traditionally had a geographically broad, if not a universal, focus. The rise of botanical gardens and zoos had been heavily intertwined with the colonial enterprise, and their collections of living organisms reflected the geographical breadth of the European empires. The scientific work performed in these places often consisted in *eliminating* the environmental variability of the colonial states, and creating homogeneous cultivation regimes for foreign animals and plants in so-called acclimatization projects. Natural history museums, likewise, were traditionally centers of accumulation, which tried to bring specimens from diverse geographical areas under a unifying classification. Laboratories, finally, strove for a complete detachment from local natural environments. By creating universally repeatable experimental set-ups and by breeding universally used purpose-bred animals, laboratory biologists separated their work from the concrete local nature outside the laboratory walls. Compared with these various places of life science, the field stations with their narrow geographical focus might seem limited in scope. Yet what was lost in spatial ambition was won in the richness of the subject matter. The interest in relations between organisms and their environment implied an eye for an almost endless number of variables. Whereas other scientific sites relied by definition on the removal of organisms from their natural habitat in order to bring them to the scientist, field stations saw much more movement back and forth. To be sure, station zoologists moved animals indoors, but many of them also used the station as a hub to delve into the field themselves. It was not isolated organisms, but nature as a whole, which they were interested in.

In its interest for local nature the research of the station zoologists echoed traditional natural history, but in other respects it broke away from it. Although zoologists such as Alfred Giard and Gustave Gilson, Otto Zacharias and Johannes Thienemann held descriptive work in high esteem, they did not see the description and classification of species and landscapes as their ultimate ambition. Through their highly detailed and localized work, they wanted to make claims that were no less universal than those made by laboratory workers. Giard wanted to reveal how parasitism worked—not only in Wimereux, but everywhere. Thienemann's counting of crows was meant to reveal the universal mechanism of migration. Zacharias hoped to unravel the circulation of matter by peering into the microcosm constituted by the Großer Plöner See. And Gilson was eager to stress that studies "even when they are limited to regional forms, directly lead to the formulation of general notions, to conclusions of a superior interest, even to new conceptions of a speculative order."[3]

To connect their local observations with universal claims, station zoologists developed a whole range of spatial strategies. Inspired by the laboratory revolution, they conceptualized nature as a lab, in which the experienced observer could monitor all kinds of "natural experiments." In these "experiments" the landscape itself performed various functions. If it was the interaction of individual species with their environment, or autecology, one was after, the landscape was often seen as an important variable. By comparing, for instance, the fauna of a brackish river with that of the sea, one could study the impact of salinity on the organisms' internal functioning. To make such a strategy work one preferably settled in an area that had a varied geological make-up and a broad range of different habitats, such as Wimereux. If the focus was on the interrelations of communities of animals, or synecology, the landscape could serve as a boundary of the microcosm one researched. In this case, the most important aspect was that it offered a relatively isolated system, such as the Großer Plöner See. In both the synecological and the autecological approach, organisms were tied (at least conceptually) to a particular landscape. Some scientists, however, showed more interest in how animals moved *through* the landscape—when and where, at which moment, speed and altitude, but also why. For such migration studies, it was particularly the density of moving animals that was helpful in choosing a site for a field station. A bottleneck in the landscape, as one encountered in Rossitten, was especially useful from that perspective. In all three cases, the landscape offered a type of controllability—an ideal that, indeed, was also cherished in a laboratory context.

Just like laboratory scientists, station zoologists often actively manipulated the natural world around them. Organisms were regularly relocated as part of outdoor experiments. Jean Massart moved plants to new habitats, while Johannes Thienemann prevented young storks from migrating with their parents. In other cases, parts of natural processes were relocated (or mimicked) indoors in order to isolate variables or to facilitate observation. Georges Bohn tried to study the reaction of invertebrates to waves in aquariums; Guillaume Severin kept insects in terrariums in order to understand their morphological variation; Thienemann studied captive storks to monitor changes in their behavior in the migratory season. All kinds of technologies were deployed to overcome nature's resistance to imprisonment in the immediate surroundings of the station. Rings and tags were used to track birds and fish, and floats had to trace the movement of the water.

All these strategies were deployed by a heterogeneous group of naturalists, whose roles were continually being redefined. To begin with, biological stations clearly were places of academic "professionalization." As indicated in chapter 1, they were material symbols that had to provide to academic zoology what university laboratories had offered physiology: authority. At the same time, however, several of them engaged amateurs of different stripes. These included the old-style gentleman scientist—who, being of independent means, could devote time, money, and energy to support field stations. Prince Albert of Monaco, founder of his own station, is a good example; Alfred Bétencourt, a regular at the station of Wimereux, another one. Yet, the late nineteenth century also witnessed the rise of a new kind of amateur. In this period, several circles started to present amateur work as a rigorous collective enterprise, rather than the occupation of sole gentleman scientists—a shift Samuel Alberti has described as "amateurization."[4] Biological field stations have played a role in both the creation and the mobilization of this new kind of amateur. Station zoologists have composed amateur field manuals (Bohn in Capus, *Guide du naturaliste*), played leading roles in new amateur societies (Rousseau in *Les naturalists belges*), and developed technologies and practices for coordinating amateur observations (Thienemann's bird-banding experiment). In most cases dealt with in this book, the eyes of teachers, landowners, fishermen, foresters, and hunters were used to monitor the landscape and the animals that lived there. Often they were regular visitors in the stations themselves. Station zoologists actively made use of the number of these people, their geographical distribution, and their particular expertise to understand the interactions between animals and the place in which they lived. Amateurs, thus, were crucial in several of the spatial strategies sketched earlier.

It was not only problems of place, however, which were dealt with in new ways by station zoologists. Problems of time were equally important. Unlike the traveler-naturalist, the station zoologist was in no hurry. He could study phenomena, at one and the same spot, over long periods. Founders of biological stations knew that long-term observation brought new scientific questions within reach. Some problems, Zacharias stressed, needed the scientist's attention "from day to day, from hour to hour."[5] His interest in yearly plankton cycles was one of these problems. Also Giard's ideas about poecilogony could be dealt with only through long and sustained examinations. The same goes for Rousseau's observations on the seasonal variation of freshwater fishes. Thienemann, for his part, was convinced that the mechanism of bird migration could be tackled only by daily counts of passing birds and daily measurements of temperature and wind direction.

In the biological station, a new sense of time and place was accompanied by a specific ideal of nature. Station zoologists such as Giard, Zacharias, Thienemann, and Gilson were interested in a nonhumanized, "unaffected" nature. Their research (unlike that of most laboratory zoologists) concerned wild animals, preferably studied in spaces that were little affected by human economies. Massart stated in 1912 that "many capital biological questions can only be studied in places where the development, the succession, the fights of animals and plants are not troubled by the intervention of man."[6] Following this ideal, biologists looked for places that bore no marks of human interference. In most of late-nineteenth- and early-twentieth-century Europe, however, nonhumanized spaces were virtually impossible to find. In the United States one might have been able to maintain the idea that a pristine wilderness was still more or less preserved behind the frontier of human civilization; in western Europe the signs of human interference were all too clear. Agriculture, forestry, and human habitation had left only small patches of wild land, and even there much of the original fauna had disappeared. Places like Wimereux, Plön, Rossitten, or Overmeire might have been natural to a certain degree, but they were hardly untouched. Even in these rather peripheral places, a centuries-old human presence was felt: the beaches and the lake regions were affected by urban tourism, the seas heavily fished, the forests exploited, the once meandering dunes now stabilized by artificial vegetation.

Station zoologists closely observed the changes that affected local species, landscapes, and habitats, and in doing so, they clearly shared a sensibility with the burgeoning nature protection movement of around 1900. In this context, it was not uncommon for zoologists to combine practices of science and nature protection. This combination was not straightforward, however, and could take different forms. Some, like Massart or Gilson, would actively propagate the protection of nature in its unspoiled state through setting up nature reserves. Part of their argument for setting up these reserves was that, through them, one would protect nature's laboratories, and thus, a major source of science itself. In other instances, station zoologists chose a more interventionist managerial approach. In many cases, they believed their role was to optimize nature's economy. For this reason, Thienemann culled foxes and introduced hole nesters, Giard hoped to spread parasites that could biologically castrate destructive insects, and Zacharias believed his research would eventually be used to increase the production of ponds. They decried some human activities—such as overfishing, the destruction of insect-eating birds, or water pollution—but not necessarily in order to protect nature in its unaltered state.

The relationship of station zoologists with the urbanizing and industrializing world was often highly ambiguous. Urban pressure pushed them to what was perceived as the last remaining patches of Europe's vanishing wilderness, but the same zoologists were heavily dependent on modern urban culture. They relied on new railways, books from urban libraries, urban professorships, industrial capital, instruments, and visitors from urban scientific centers. They were aware that urbanization stimulated intensive farming, forestry, and fisheries and in this way threatened the scraps of (semi-wild) nature crucial for their work. Yet at the same time they stressed that their studies of animals in their natural habitat would optimize intensive forms of exploiting nature. They dreaded tourists in the vicinity of their station, but their own depictions of unspoiled landscapes attracted ever more urban holiday-makers.

Despite its material and cultural reliance on the city, the place-based research in Europe was a reclusive project. The zoological station was often allegorically construed as a home in isolation, a hermitage, an outpost in the desert. This image was associated with a particular rhetoric of domesticity, in which the station director was the patriarch of a scientific family. Often the rhetoric of domestic retreat was matched with an equally hermitlike discourse of scarcity, in which the patriarch managed his household as economically as possible. Life in the proto-ecological station was presented as being simple, and, although the limited means and equipment were a constant source of complaint, station directors carried a certain pride in the paucity of their working conditions. The improvised character of both the stations and the science performed there, was explicitly cultivated. The small chalet in Wimereux, the old painter's studio in Rossitten, the quickly altering locations of Gilson's laboratory in Ostend, all became places of emotional attachment and symbols of a science that owed more to perseverance and patient observation than to expensive laboratory equipment. The station in Plön, then, might have been villa-like, but Zacharias made sure to stress that it certainly was no "imposing temple of science."[7] Obviously, the emphasis on sobriety and lack of equipment (together with hints at foreign luxury) was part of a fundraising strategy. Yet the rhetoric of scarcity also became part of the self-image of many station zoologists. It matched well with the anti-technological stance of Giard, the self-claimed soberness of Zacharias, the hunter's romanticism of Thienemann, and the seaman's heroics of Gilson.

The self-presentation of the station zoologists helped create space for a new scientific perspective. The topoi of isolation and adventure, the stress on patience and practical improvising skills, were after all closely associated with the program of studying the interaction between animals and their environment. They coincided with counter-caricatures of urban laboratory-based science, which was purportedly luxuriously housed, easy to perform, predictable, and narrow in its perspective. Place-based station zoology defended different values, because it asked different questions.

Late-nineteenth- and early-twentieth-century station zoologists created a conceptual space for asking (proto-)ecological questions, and generated the technological and conceptual tools to tackle these. Obviously, station zoologists were not the only actors to stir ecological awareness at the turn of the nineteenth century. The interaction of animals with each other and their environment was an interest that could be found among practical naturalists and popular science writers, and it was also developed by a limited number of elite scientists who were working outside the stations' context. It is the latter figures, scientists such as Karl Möbius in the late nineteenth century and Charles Elton in the early twentieth, who have been best remembered by ecologists and historians alike. Far from working in an intellectual vacuum, however, these men were picking up on a broad range of ideas and practices of which the zoological stations were crucial strongholds.

The case studies discussed in this book deal with early examples of such strongholds. Later, comparable stations would promote similar ideas and practices in other national and regional contexts. Ecological traditions developed among others in the station of Blakeney Point in the United Kingdom (founded in 1913), Lough Ine in Ireland (1925), Wijster in the Netherlands (1927), and Bellinchen in Germany (1928). The ecological sensibilities that developed in these places were largely interchangeable, but the institutional role played by zoological stations differed from one national context to the next. In the Anglo-American world, ecology turned into a fully fledged discipline, from the 1910s onward, with its own societies, journals, and university chairs. In countries such as France and Germany similar developments only materialized four to five decades later, which made stations the most important loci around which ecologically inspired research could cluster in the late nineteenth century and the first half of the twentieth.[8] It is in this clustering that much of their scientific significance has to be sought.

Stations brought together people from different backgrounds and acted as trading zones where these people could find a common language. This enabled the transfer of knowledge between various groups. In German and Francophone science, biological stations particularly served as transit points between the world of practical naturalists and nature writers on the one hand and the elite culture of professional scientists on the other. In the French-speaking world, the transit of ecological knowledge largely had a top-down character. It was elite scientists such as Giard and Gilson who organized the station, set up its scientific program, and controlled its social structure. At the same time, they involved amateurs, students, fishers, and farmers in their projects, and made sure to propagate their ecologically inspired ideas outside the world of professional science. In Germany, by contrast, the zoological station rather served as a passage point of ideas and practices that moved from the bottom up. Plön was explicitly created to transfer a somewhat Humboldtian interest in freshwater animals from the world of amateur aquarists and science journalists to the zoology departments at the universities. In Rossitten, then, ornithological questions that previously had largely been dealt with by amateurs, hunter-naturalists, and nature writers, would slowly find their way to professional zoologists. The integration of both stations into the Kaiser Wilhelm Society was indicative of the track ecological interest followed in these places.

Despite these national variances, the proto-ecological projects of the zoological stations had similar consequences for German and Francophone science. They professionalized the interest in the interactions of animals with each other and their environment; they echoed sensibilities of the laboratory in the field, but extended (and partially transformed) the laboratory's focus and interests; they made the vestiges of "unspoiled" European nature into crucial scientific sites and used these sites to understand the natural world as a whole and to manage its resources.

Biological stations generated ideas and practices that were deemed important beyond the specific context in which they originated. One reason was that place-based research contained the promise of (eventual) economic application. The study of freshwater plankton was believed to offer a key to increasing the production of ponds, research into invertebrate parasites was presented as instrumental in modernizing both the fisheries and agriculture, and the analysis of bird habits was seen as helpful in the battle against insect pests. But the study of animals in interaction with their environment was not just a utilitarian enterprise; it was also in line with broader cultural ideals. In the Francophone world, a sturdy scientism and environmental determinism was central in creating support for the station movement. Understanding how individual organisms (of both animals and men) interact with their environment became one of the central interests of the Third Republic. In the German Reich, then, station directors promoted their research as central to valuing the *Heimat* and as a crucial instrument for personal *Bildung*. By playing into these sensibilities station zoologists were able to connect with society and spread the idea that organisms could be understood only in interaction. In order to get this message across, they made sure to publicize their research beyond the small world of specialists. They actively translated their findings to fishery, hunting, and forestry journals, museum exhibits, nature walks, popular books and films, field guides, and courses for schoolteachers.

The project of studying animals in relation to their environment by making the zoologist a resident *in* this environment was not without its ambiguities. It was a project that strove for an experimental aura, while at the same time downplaying the importance of scientific manipulations and other "artifice." It was a project of retreat from the city that nevertheless depended on urban technologies and urban intellectual input. It was an inclusive enterprise that involved many nonscientists, but that also carried strong professional (or professionalizing) ambitions. It was a project of pure science, of which the proponents proclaimed its economic significance. It was, finally, a project that aimed at universal truths by revaluing the local.

The science performed in zoological stations was of an ambiguous kind indeed. Therefore I do not think we can understand the station movement by using a simple dichotomy between the lab and the field, in which the rise of the zoological stations should be seen as a colonization of the field by laboratory technologies and methods. Neither can we understand it by using simple dichotomies between nature and non-nature, in which the station movement would have been about the rediscovery of the first. To study the interactions between animals and their environment, the station zoologists relied on mixed practices that indeed borrowed partially from laboratory traditions, but also from those developed in several other scientific workplaces. Their research, furthermore, was never performed in isolation in an eternal wild nature; they always relied on material and intellectual connections to the rapidly changing outside world, which also provided them with the rationale for their work. What nature actually was, was intellectually constructed in the process.

Acknowledgments

This book started as a research project at the University of Leuven, was developed during two visiting scholarships at the University of Cambridge and Imperial College London, respectively, and was finished at the University of Maastricht. All the while it was generously sponsored by the Research Foundation-Flanders.

Richard Burkhardt, Josquin Debaz, Nick Hopwood, Laurent Loison, Liesbet Nys, Anne Secord, Fabio de Sio, Astrid Schwarz, Harro Strehlow, Megan Raby, Christian Reiss, François Schmitt, Robert-Jan Wille, and Kaat Wils helped with suggestions and advice. Keith Benson, Vanessa Heggie, Jesse Olszynko-Gryn, Helen MacDonald, Anita McConnell, Andrew Mendelsohn, Marion Thomas, and Jo Tollebeek read drafts of book chapters at various stages of completion, and offered stimulating comments. I owe a particular debt to Lynn Nyhart, who read the whole manuscript with care and commented constructively and with expertise. Three anonymous referees provided generous advice. Special thanks go to Karen Merikangas Darling, at the University of Chicago Press, for her interest in this project and her unstinting support to turn it into an actual book. Obviously (as readers of acknowledgments know), the remaining errors in this book are entirely mine.

For archival material I relied on the personnel of the Wellcome Library and the Imperial College Library in London, the University Library in Cambridge, the Bodleian Library in Oxford, the Institut Pasteur in Paris, the Belgian Institute for the Natural Sciences and the State Archives in Brussels, the university archives of Liège, Brussels, and Louvain-la-Neuve, the Stazione Zoologica in Naples, the Universitätsbibliothek and the Zoologisches Museum in Kiel, the Max-Planck-Institut für Ornithologie in Radolfzell (where the study tables have a view on overflying storks), the Max-Planck-Institut für Evolutionsbiologie in Plön, the Bundesarchiv-Filmarchiv and the Staatsbibliothek in Berlin. Particular thanks go to Rolf Schlenker for his warm welcome in Radolfzell and for the access he gave to his wonderful personal archive collections.

Parts of the argument developed in this book go back to earlier articles. Chapter 6 draws on an article published in *Journal of the History of Biology* 44, no. 2 (May 2011), under the title "Poetry and Precision: Johannes Thienemann, the Bird Observatory in Rossitten and Civic Ornithology, 1900–1930." I thank Springer for granting me permission to republish parts of this article here. Parts of other chapters deal with material I touched upon in articles in *Social Studies of Science* (in 2009) and *Isis* (in 2010). These have been substantially rewritten for the book.

That I will have lasting good memories of writing this book has to be ascribed to the habitat in which it was written. In the good memories, Greet, Nand, and Linus take a central place.

Finally I want to thank my parents, who have always stimulated my interest for all things historical. This book is dedicated to them.

Notes

INTRODUCTION

1. Zacharias, "Die modern Hydrobiologie" (1905), 85.
2. For the expression "engines of experiment": Dierig, "Engines for Experiment" (2003).
3. Billick and Price, "The Ecology of Place" (2010), 4-6.
4. Jack, *The Biological Field Stations* (1940); idem, "Biological Field Stations" (1945).
5. The recent volume of *The Cambridge History of Science* dedicated to the modern biological and earth sciences contains an informative article by Keith Benson that summarizes the historical literature on "field stations and surveys." Tellingly, it is the shortest chapter of the volume. Benson, "Field Stations and Surveys" (2009).
6. The notable exceptions are Edward Eigen's "The Place of Distribution" (2003) and Jeremy Vetter's "Rocky Mountain High Science" (2010), next to important passages in Robert Kohler's *Landscapes and Labscapes* (2002).
7. See among others, Groeben, "Anton Dohrn" (1985); Pauly, "Summer Resort" (1988); Maienschein, *100 Years Exploring Life* (1989); Pauly, *Biologists* (2000), 145-164; Erlingsson, "The Plymouth Laboratory" (2009).
8. Benson, "Laboratories on the New England Shore" (1988); Kohler, *Landscapes and Labscapes* (2002), 42-44.
9. Allen, *The Naturalist in Britain* (1994),192.
10. See: Tobey, *Saving the Prairies* (1981); Worster, *Nature's Economy* (1985), 192-219; Söderqvist, *The Ecologists* (1986); special issue "Reflections on Ecology and Evolution" of *Journal of the History of Biology* (1986); Cittadino, *Nature as the Laboratory* (1990); Bocking, *Ecologists* (1997); Acot, *The European Origins of Scientific Ecology* (1998); Kingsland, *The Evolution of American Ecology* (2005).
11. Juday, "Some European Biological Stations" (1910), 1258.
12. Elton, *Animal Ecology* (1927), 3. Historians have stuck to Elton's explanation. See, e.g., Barrow, *Nature's Ghosts* (2009), 207.
13. Billick and Price, "The Ecology of Place" (2010), 6.
14. See among others, Pickstone, *Ways of Knowing* (2000), 135-161; Bowler and Rhys Morus, *Making Modern Science* (2005), 165-182; Daston and Lunbeck, "Introduction" (2011), 3-5.
15. Cittadino, *Nature as the Laboratory* (1990).
16. Wille, "The Co-production" (forthcoming).
17. Raby, *Making Biology Tropical* (2013).

18. Kohler, *Landscapes and Labscapes* (2002), 76.

19. E.g., Ray, "The Application of Science" (1979), 250–251; Reingold and Reingold, *Science in America* (1981), 2–6; Bruce, *The Launching of Modern American Science* (1987), 64–74; Timmons, *Science and Technology* (2005), 2–5.

20. See particularly Nash, *Wilderness* (1967).

21. Worster, *Nature's Economy* (1985), 241.

22. Nyhart, *Modern Nature* (2009), 2.

23. Alberti, "Amateurs and Professionals" (2001); Allen, "Amateurs and Professionals" (2009); Vetter, "Introduction: Lay Participation" (2011).

24. Galison, *Image and Logic* (1997), 803–844.

25. See among others, Golinski, *Making Natural Knowledge* (1998), 79–102; Smith and Agar, *Making Space for Science* (1998); Kohler, "Place and Practice" (2002); Livingstone, *Putting Science in Its Place* (2003); Eigen, "The Place of Distribution" (2003); Mitman, Murphy, and Sellers, *Landscapes of Exposure* (2004); Naylor, "Introduction" (2005); Finnegan, "The Spatial Turn" (2008); Kingsland, "The Role of Place" (2010).

26. More generally on the relationship between science and urbanity: Eckart, *Wissenschaft und Stadt* (1992), 69–225; Dierig, Lachmund, and Mendelsohn, *Science in the City* (2003). On the urban aspects of field excursions, see: Phillips, "Friends of Nature" (2003).

27. For these developments see among others: Doughty, *Feather Fashions* (1975); Zirnstein, *Ökologie und Umwelt* (1994), 95–14; Büschenfeld, *Flüsse und Kloaken* (1997); Gudermann, "Der Take-off der Landwirtschaft" (2001); Jansen, "*Schädlinge*" (2003); Pitte, *Histoire du paysage français* (2003), 267–303; Blackbourn, *The Conquest of Nature* (2006).

28. Giard, "Les tendances actuelles" (1905), 134.

CHAPTER ONE

1. Later sources would backdate the foundation to 1842. Sand, "Les laboratoires maritimes" (1898), 23.

2. Sand, "Les laboratoires maritimes" (1898), 126–127.

3. Van Beneden, "Recherches sur la structure de l'oeuf " (1841), 90.

4. Hamoir, *La révolution évolutioniste* (2002), 13–14.

5. Secord, "Coming to Attention" (2011). On the "seaweed craze": Allen, "Tastes and Crazes" (1996), 397–400. Although the country house became less prominent as a site of research throughout the nineteenth century, country-house science did not disappear and—particularly in the United Kingdom—underwent its own kind of evolution. See: Opitz, "'This House Is a Temple'" (2006).

6. Allen, *The Naturalist in Britain* (1994), 111–117; Florey, "Highlights and Sidelights" (1995), 89–83; Eigen, "The Place of Distribution" (2003), 60–68.

7. Lenoir, "Laboratories, Medicine and Public Life" (1992), 68–69.

8. Johannes Müller to Max Müller, 6 Nov. 1848, in: Haberling, *Johannes Müller* (1924), 324.

9. Koller, *Johannes Müller* (1958), 184.

10. See among others, Lacaze-Duthiers, "Direction des études"(1872), 47 and 62; Giard, "Particularités" (1911 [1878]), 512–513; Groeben et al., *Emil du Bois-Reymond* (1985), 253; Heuss, *Anton Dohrn* (1991), 169.

11. Next to Müller, Van Beneden, and Milne-Edwards, these men included Louis Agassiz, Sven Lovén, Johan Japetus Steenstrup, Michael Sars, Albert von Kölliker, Christian Ehrenberg, and Carl von Siebold. See among others, Winsor, *Starfish, Jellyfish* (1976); Rehbock, "The Early Dredgers" (1979), 296; Allen, *The Naturalist in Britain* (1994), 111–117.

12. Otis, *Müller's Lab* (2007), 26–31; Van Dyck, "De zoölogische verzamelingen" (1983).

13. Kemna, *P. J. Van Beneden* (1897), 14.

14. De Fonvielle, "Coste" (1873); Fauré-Fremiet, *Notice sur la vie et les travaux de Victor Coste* (1960); Baguley, *Napoleon III* (2000), 195-206; Kinsey, "'Seeding the Water'" (2006), 536-537.

15. Coste, "Mémoire sur les moyens de repeupler les eaux" (1853).

16. Debaz, *Les stations françaises* (2005), 119-124; Pavé, "'To Capture or Not to Capture'" (2005); Kinsey, "'Seeding the Water'" (2006), 532-538.

17. Coste, *De l'observation et de l'expérience* (1869), 23-25.

18. De Quatrefages, *Fertilité et culture d'eau* (1862), 11.

19. Kinsey, "'Seeding the Water'" (2006), 539 and 558-559.

20. The archives of the Wellcome Trust in London contain a peculiar scrapbook that collects disapproving material on Coste. Next to caricatures meticulously copied from *Le Charivari*, it encloses derogatory songs, critical articles, and letters of complaint. The collection is anonymous, but clearly made by a high-ranking maritime administrator. The letter cited was written by an unnamed commissar-general of the maritime administration, 28 May 1862. Wellcome Trust, WMS 4 MS 1888.

21. To be sure, the history of the station of Arcachon dates back to 1867, but in origin this was only an aquarium and a museum. The laboratory connected to it opened its doors only in 1886.

22. Some leading station directors were antagonistic toward Darwinism (such as Coste) or noncommittal (such as Lacaze-Duthiers), but also those who were staunch Darwinians did not present evolution as the only area of interest. See: Stebbins, "France" (1988), 138-139 and 148-149.

23. On this spirit, see among others: Barral, *Les Fondateurs* (1968); Anderson, *France 1870-1914* (1977).

24. It would, overall, be fair to typify him as a conservative republican. Lacaze-Duthiers, "Création d'un laboratoire de zoologie" (1872), lii; idem, "Dix-sept années" (1886), 741. For his biography: De Parville, "Henri de Lacaze-Duthiers" (1901); Appel, "Lacaze-Duthiers" (1973); Bouyssi, *Alfred Giard* (1999), 102-103.

25. Lacaze-Duthiers to Léon Frédericq, 5 March 1888, in: Florkin and Théodorides, *Henri de Lacaze-Duthiers* (1982), 52.

26. For more details see among others: Kofoid, *The Biological Stations* (1910), 35-143; Caullery, "Les stations françaises" (1950); Paul, *From Knowledge to Power* (1985), 103-117.

27. Groeben, "Tourists in Science" (2008). On Goethe's *Italienfahrt*: Richards, *The Romantic Conception* (2002), 382-409.

28. Haeckel went so far as to "re-enact" Goethe's climbing of the Vesuvius. See Richards, *The Tragic Sense of Life* (2007), 55-63.

29. In a letter to Darwin, Dohrn refers to a Messina fisherman who attended Müller, Kölliker, Gegenbaur, Haeckel, Claus, and himself. Dohrn to Darwin, 30 Dec. 1869, in: Groeben, *Charles Darwin* (1982), 27.

30. The quote is from the Italian zoologist Filippo de Filippi and is cited in: Heuss, *Anton Dohrn* (1991), 75. *Privatdozent* means private lecturer.

31. Strehlow, "Zur Geschichte des Berliner Aquariums" (1987).

32. The Stazione zoologica will be dealt with in more depth in chapter 2.

33. Müller, "Der 'Hydriot' Nikolai Kleinenberg" (1973).

34. "Marine Zoological Laboratories" (1883), 16-17.

35. This was among others the case in the Norwegian stations of Flödevig (founded in 1880) and Trondheim (1890), and in the British stations of Dunbar (1880), Aberdeen (1890), and Piel-in-Barrow (1896).

36. Kofoid, *The Biological Stations* (1910), 144-213 and 279-316.

37. Ehrenbaum, "Zoologische Wanderstation" (1889); Heincke, "Die Biologische Anstalt" (1893); Kofoid, *The Biological Stations* (1910), 221-222; Werner, *Die Gründung* (1993), 6-26.

38. Lacaze-Duthiers, "Avertissement" (1872), v-vi.

39. Lacaze-Duthiers touched on all these elements when he informed his friend Thomas Huxley about his newly established station in Roscoff: "I hope that within the year, the work and the workers will be installed here; without luxury in apparatus, without big and immense aquariums for the public, but with the necessary things for study. I am full of confidence in the future of my country, which I love." Lacaze-Duthiers to Huxley, 12 August 1872, Huxley Correspondence, Imperial College London.

40. On the Republic's scientific aura, among others: Weisz, *The Emergence of Modern Universities* (1983), 90-161; Fox, "Science" (1984), 101-118; Anderson, *European Universities* (2004), 176-190.

41. Debaz, *Les stations françaises* (2005), 117.

42. Half of the delegates in the Bureau of the Association were marine zoologists, and fifty percent of the budget of the section would be spent on the construction of marine laboratories or projects carried out there. Van-Praët, "La section Zoologie" (2002), 159-160.

43. De Parville, "Henri de Lacaze-Duthiers" (1901), 139-140; Kofoid, *The Biological Stations* (1910), passim; Giard, "Discours" (1913 [1874]) 8; Paul, *From Knowledge to Power* (1985), 109.

44. This was particularly so for zoology chairs. Between 1880 and 1910 their total number barely increased from 26 to 28. Harwood, "National Styles" (1987), 398.

45. Busch, "The Vicissitudes of the *Privatdozent*" (1962); Ben-David, "The Universities" (1974), 46-50; Anderson, *European Universities* (2004), 151-161.

46. On Heincke: Werner, *Die Gründung*, 28-30.

47. Heuss, *Anton Dohrn* (1991), 163.

48. Schwarz, *Frühe Ökologie* (2000), 124-129.

49. On the early years of limnology, see: Zacharias, *Das Süsswasser-Plankton* (1907), 4-7; Steleanu, *Geschichte der Limnologie* (1989), 186-269. For a more up-to-date (and a more philosophical) approach: Schwarz, *Wasserwüste* (2003).

50. Kofoid, *The Biological Stations* (1910), 233-245 and 266-272; Juday, "Some European Biological Stations" (1910), 1273-1274; Lenz, "Limnologische Laboratorien" (1936), 1296-1311.

51. Harwood, "Weimar Culture" (1996), 348-351.

52. Girod, "La station biologique" (1893); Bruyant, "La station limnologique" (1900); Reynouard, "Besse d'aujourd'hui" (1909); Kofoid, *The Biological Stations* (1910),134-143.

53. Lestage, "Le Dr. Ernest Rousseau" (1921); Lameere, "L'Institut zoologique" (1927).

54. Egerton, "Ecological Studies" (1976), 339-340; McIntosh, *The Background of Ecology* (1985), 59-60.

55. Zacharias, "Über die systematische Durchforschung" (1905), 23-24.

56. Lenz, "Limnologische Laboratorien"(1936), 1287.

57. Jansen, "Den Heringen" (2002).

58. Von Lucanus, *Die Rätsel des Vogelzuges* (1922), 14-16; Stresemann, *Die Entwicklung* (1951), 338-339; Allen, *The Naturalist in Britain* (1994), 198-200.

59. Stresemann, *Die Entwicklung* (1951), 341-343.

60. Von Lucanus, *Die Rätsel des Vogelzuges* (1922), 22-24.

61. One can get a good idea of the composition of this community from the group images of the yearly gatherings of the German Ornithological Society: Haffer, "Gruppenbilder" (2003).

62. Despite the commonly acknowledged importance of the laboratory there does not yet exist a general historical overview on the subject: Kohler, "Lab History" (2008). About the interconnected rise of the laboratory and experimentalism, see Pickstone, *Ways of Knowing* (2000), 135-161.

63. Kohler, *Landscapes and Labscapes* (2002), 3.

64. Minot, "The Laboratory" (1884), 172-174.

65. Thomson, "Scientific Laboratories" (1885), 411-412.

66. Galison, "Buildings" (1999), 1.

67. The quotation comes from: Bernard, *An Introduction* (1957 [1865]), 146. For the history of midcentury physiology: Lenoir, *The Strategy of Life* (1982), 195–245; Lesch, *Science and Medicine* (1984), 197–224; Coleman, "The Cognitive Basis" (1985); Coleman and Holmes, *The Investigative Enterprise* (1988); Lenoir, "Laboratories, Medicine and Public Life" (1992); Kremer, "Building Institutes" (1992); Dierig, "Engines for Experiment" (2003); idem, *Wissenschaft in der Maschinenstadt* (2006).

68. E.g., for a sketch of Ludwig's institute in 1858, see: Ludwig to Emil Du Bois-Reymond, March 14, 1858, in: Estelle Du Bois-Reymond, *Two Great Scientists* (1982), 96–97.

69. A lively description of du Bois-Reymond's early laboratories can be found in: Dierig, *Wissenschaft in der Maschinenstadt* (2006), 26–30 and 49–55.

70. Fox, "Scientific Enterprise" (1976), 24–29; Miles, "Reports by Louis Pasteur" (1982–1983), 116.

71. Welch, "The Evolution" (1896), 88.

72. Nyhart, *Biology Takes Form* (1995), 90–91.

73. Paul, *From Knowledge to Power* (1985), 93.

74. As such there was a clear difference of opinion with the earlier generation. See, e.g., Emil Du Bois-Reymond to Ludwig, August 7, 1849, in: Estelle Du Bois-Reymond, *Two Great Scientists* (1982), 46.

75. Notably in: Paul, *From Knowledge to Power* (1985), 94–103; Maienschein, "Arguments for Experimentation" (1986), 181–182; Marinescu and Bratescu, "Une controverse" (1992); Debaz, *Les stations françaises* (2005), 361–369.

76. Bernard, *Rapport* (1867), 131–148; Bernard, *Notes* (1979), 58. The quoted notes were written in preparation of the 1867 report but were published only posthumously in 1979.

77. This did not prevent Bernard from repeating them in equally strong tones in the following years. E.g., Bernard, "Médicine expérimentale" (1868).

78. Coste, *De l'observation et de l'expérience* (1869), 5–8 and 23.

79. See: Chevreul, *De la méthode* (1870).

80. Of the original article Lacaze-Duthiers' "Direction des études"(1872), large parts are translated into English for Coleman's anthology of nineteenth-century morphological texts. My quotations come from this translation. Lacaze-Duthiers, "The Study of Zoology" (1967 [1872]), 135–140 and 145.

81. E.g., Giard, "Les tendances actuelles" (1905), 29–30.

82. Nyhart, *Biology Takes Form* (1995), 169–175.

83. See Dohrn, *Kurzer Abriss* (1871); Dohrn to du Bois-Reymond 2 Aug. 1871, 18 Sept. 1871, and 28 March 1873, and du Bois-Reymond to Dohrn 4 Aug. 1871, in Groeben et al., *Emil du Bois-Reymond* (1985), 1–2, 5–7, and 47.

84. Du Bois-Reymond would also act in support of the freshwater station in Plön and, despite Dohrn's objections, he would eventually approve of the station in Helgoland. Du Bois-Reymond to Zacharias, 24 July 1893, Universitätsbibliothek Kiel, Nachlass Zacharias; Heuss, *Anton Dohrn* (1991), 167; Werner, *Die Gründung* (1993), 11 and 17.

85. Nyhart, "Natural History" (1996), 435.

86. Shinn, "The French Science Faculty System" (1979), 305.

87. De Bont, "Evolutionary Morphology" (2008), 101–102.

88. Naples was known for its trade in specimens, but also other stations developed a lively activity in this regard. Lacaze-Duthiers, for example, prided himself on the fact that his laboratories in Roscoff and Banyuls took care of 240 transports of living animals throughout France in the academic year 1885–1886 alone. Lacaze-Duthiers, "Dix-sept années" (1886), 742.

89. More extensively on the glass case and the Victorian aquarium craze: Lloyd, "Aquaria" (1876); Rehbock, "The Victorian Aquarium" (1980); Hamlin, "Robert Warington" (1986); Brunner, *The Ocean at Home* (2003), 30–58; Darby, "Unnatural History" (2007).

90. Lloyd, "Aquaria" (1876), 260; Brunner, *The Ocean at Home* (2003), 99-120.

91. Bashford, "Public Aquariums" (1896), 17.

92. As Celeste Olalquiaga has shown, Lloyd was far from pleased with the increasing kitschiness of public aquariums. Olalquiaga, *The Artificial Kingdom* (1999), 154-156.

93. At several instances Dohrn recounted this "eureka moment" of his visit to the Berlin Aquarium. His historians have taken over the story. Dohrn, "Der gegenwartige Stand" (1926 [1872]), 417; Heuss, *Anton Dohrn* (1991), 112-114, 150-151, and 267.

94. Prouho, "Le laboratoire Arago" (1886); Kofoid, *The Biological Stations* (1910), 72-75. The quote comes from: Boutan, *La photographie* (1900), 116.

95. Dohrn, *Erster Jahresbericht* (1876), 2; idem, "Bericht über die Zoologische Station" (1879), 141-143; Prouho, "Le laboratoire Arago" (1886), 98; Lacaze-Duthiers, "Sur les laboratoires" (1898) 21; Kofoid, *The Biological Stations* (1910), passim.

96. Du Bois-Reymond, "Lebende Zitterrochen in Berlin" (1885), 691; Guttstadt, *Die naturwissenschaftlichen und medizinischen Staatsanstalten* (1886), 246-247; Schmeil, "Die zoologische Station" (1893); Bashford, "Public Aquariums" (1896), 25-26; Strehlow, "Zur Geschichte des Berliner Aquariums" (1987); Klös et al., *Die Arche Noah* (1994), 346-353.

97. Boutan, "Mémoire sur la photographie" (1893); idem, *La photographie* (1900); Eigen, "Dark Space" (2001).

98. On the image of the necropolis, launched by Edmond Perrier, see: Eigen, "The Place of Distribution" (2003), 57.

99. Perrier—zoologist at the Muséum—opened the article that announced the foundation of his station with the bitter words: "It has been a long-time fashion to speak ill of the Muséum d'Histoire Naturelle, and sometimes one still speaks ill of it today." Perrier, "Le laboratoire maritime" (1888), 186.

100. See more extensively in chapter 7.

101. Good examples include the museums of Arcachon, Endoume, Sète, Besse, and Rossitten.

102. "The Oceanographical Museum" (1910), 192.

103. Richard, *Le Musée océanographique* (n.d.), 9-52.

104. Kofoid, *The Biological Stations* (1910), 228.

105. On the metropolitan developments: Wonders, *Habitat Dioramas* (1993); Nyhart, "Science, Art and Authenticity" (2004); Kretschmann, *Räume öffnen sich* (2006), 76-86.

106. On the late-nineteenth-century revival of naturalist societies: Fox, "The Savant Confronts His Peers" (1980); Matagne, *Aux origines de l'écologie* (1999), 43-53; Daum, *Wissenschaftspopularisierung* (1998), 104-109; De Bont, *Darwins Kleinkinderen* (2008), 201; Nyhart, *Modern Nature* (2009), 41.

107. On these excursionist instruments: Larsen, "Equipment for the Field" (1996).

108. Lacaze-Duthiers, "Sur les laboratoires" (1898), 3.

109. Zacharias, "Vorwort" (1896), vi.

110. Sauvage, "La station zoologique" (1883).

111. Cunningham, "La station maritime" (1884), xviii.

112. See respectively: Frič and Vavrá, *Die Thierwelt* (1894), 8-13; "Nouvelles scientifiques" (1891), 85-87; Lauterborn, "Das Projekt eine Schwimmende Biologische Station" (1902).

113. Ruttner, "Ein mobiles Laboratorium" (1933).

114. This argument has particularly been worked out for eighteenth-century expedition ships: Sorrenson, "The Ship as a Scientific Instrument" (1996).

115. On the history of oceanography for the period discussed: Mills, *Biological Oceanography* (1989), 1-186; Deacon, *Scientists and the Sea* (1997), 306-406; Rozwadowski, *Fathoming the Ocean* (2005), 135-208; Nyhart, "Voyaging" (2012).

116. Kofoid, *The Biological Stations* (1910), 38-48. For more on the scientific enterprise of Albert I and its role in French oceanography: Mills, *The Fluid Envelope* (2009), 162-191.

117. Werner, *Die Gründung* (1993), 100-115.

118. Groeben, "The Vettor Pisani Circumnavigation" (1990).

119. For an overview of limnological and "nautical" equipment in the late nineteenth century, see respectively: *Handbuch der nautischen Instrumente* (1882); Frič and Vavrá, *Die Thierwelt* (1894), 12; For the transfer of Hensen's methodology: Zacharias, "Quantitative Untersuchungen" (1896).

120. Rozwadowski, *Fathoming the Ocean* (2005), 183.

121. The last report of the Challenger expedition only appeared eighteen years after its return. Hensen, for his part, would bring together the conclusions of the Plankton-expedition only in 1911, twenty-two years after his expedition ship returned to the harbor of Kiel.

122. Kofoid, *The Biological Stations* (1910), 30.

123. On this promise: Tucker, *Nature Exposed* (2005); Daston and Galison, *Objectivity* (2010), 125-138.

124. See respectively: Journal A. Note 1 à 311 (1905-1906), Archives of the Institut Pasteur, Fonds Casimir Cépède, 183; Photographs, RBINS, Severin Papers Carnets; Massart, *Pour la protection* (1912).

125. Microphotography was taken up very early at the station of Algiers, and later also in Plön, Endoume, Saint-Vaast-la-Hougue, and the Royal Hungarian Biological Station for Fisheries. Aerial photography was particularly defended by the geologist and limnologist Erich Wasmund, who was subsequently attached to the biogeological station of Wasserburg am Bodensee and the limnological station in Plön. See: Viguier, "La photographie" (1888); Zacharias, "Das Plankton" (1905), 297; Kofoid, *The Biological Stations* (1910), 56, 111, 278; Wasmund, "Luftfahrzeuge" (1929). On Wasmund's combination of aerial photography and limnology (and its limited success): Schwarz, "Rising above the Horizon" (2009).

126. Lacaze-Duthiers, "Sur les laboratoires" (1898), 21.

CHAPTER TWO

1. Edwards, "The Zoological Station" (1910), 225.

2. See for example: Whitman, "The Advantages" (1883), 94; Herdman, "The Greatest Biological Station" (1901), 429; Kofoid, *The Biological Stations* (1910), 9; Williams, *A History* (1912), vol. 5, 121.

3. See among others: Müller, "Die Wandlung" (1975); Monroy and Groeben, "The 'New' Embryology" (1985); Maienschein, "First Impressions" (1986); Müller, "The Impact" (1996); Fantini, "The 'Stazione Zoologica'" (2000).

4. Dohrn, "Der gegenwärtige Stand" (1926 [1872]); Dohrn, "The Foundation" (1872).

5. Anton Dohrn to Fanny Lewald and Adolf Stahr, 19 April 1866, quoted in Groeben, "Anton Dohrn" (1985), 5.

6. For a book-length biography: Heuss, *Anton Dohrn* (1991).

7. Dohrn, "Der gegenwärtige Stand" (1926 [1872]), 416.

8. Dohrn to du Bois-Reymond, 17 May 1883, Groeben et al., *Emil du Bois-Reymond* (1985), 246.

9. With regard to the development of Dohrn's scientific ideas, see: Dohrn, *Der Ursprung der Wirbelthiere* (1875); Bowler, *Life's Splendid Drama* (1996), 114-115 and 160-167. More in particular for his relation with Gegenbaur and Haeckel: Ghiselin, "Carl Gegenbaur" (2003); Di Gregorio, *From Here to Eternity* (2005), 324-337. A good overview of Dohrn's activities in the 1860s can be found in: Groeben, "The Stazione" (2006).

10. Dohrn, "The Foundation" (1872), 277 and 279.

11. The quotes are from: Dohrn, "Der gegenwärtige Stand" (1926 [1872]), 419; Dohrn, "The Foundation" (1872), 279.

12. The same historians have shown that phylogenetic embryology was, by the 1890s, gradually replaced by the "new" experimental embryology, of which the Naples station would also become an important stronghold. For the phylogenetic work in Naples, see among others: Maienschein, "It's a Long Way from *Amphioxus*" (1994); Ghiselin and Groeben, "Elias Metschnikoff" (1997); Bowler, *Life's Splendid Drama* (1996), 114-115 and 160-167 . For the later rise of new embryology: Monroy and Groeben, "The 'New' Embryology" (1985).

13. The quotes come from: Dohrn, "Der gegenwärtige Stand" (1926 [1872]), 417-418; Dohrn, "The Foundation" (1872), 279.

14. Dohrn is referring to Adelbert von Chamisso, a Prussian poet and botanist, and Johann Dzierzon, a Silesian priest known for his studies of bees.

15. For the quotes: Dohrn, "The Foundation" (1872), 279. See also: Dohrn, "Der gegenwärtige Stand" (1926 [1872]), 417-419.

16. Dohrn to Adolf Stahr and Fanny Lewald, 15 April 1866, quoted in: Di Gregorio, *From Here to Eternity* (2005), 326.

17. Helmholtz et al., "Die zoologische station" (1980 [1879]), 49.

18. Astarita, *Between Salt Water and Holy Water* (2005), 220-275.

19. Groeben, *Charles Darwin* (1982), 33, 57-58, 95-96.

20. Snowden, *Naples* (1995), 17.

21. Baedeker, *Italien: Handbuch* (1880), 81-82.

22. Nunn Whitman, "The Zoological Station" (1886), 791.

23. See, e.g., Dohrn to du Bois-Reymond, 23 April 1876 and 26 March 1882, in Groeben et al., *Emil du Bois-Reymond* (1985), 85 and 234.

24. Snowden, *Naples* (1995), 11-30.

25. E.g., Dohrn to von Baer, 18 August 1875, Groeben, *Correspondence* (1993), 139; Dohrn to du Bois-Reymond, 24 Jan. 1880 and 14 Oct. 1886, Groeben et al., *Emil du Bois-Reymond* (1985), 183 and 264; Dohrn to Virchow, 13 Oct. 1886, Groeben and Wenig, *Anton Dohrn* (1992), 69.

26. The quote comes from a text by the British naturalist, science popularizer, and tide pool enthusiast Philip Henry Gosse, but reflects a wider practice both in the UK and the European Continent. Gosse, "Sea-side Recreations" (1853), 298.

27. Snowden, *Naples* (1995).

28. Kofoid, *The Biological Stations* (1910), 31.

29. Dohrn, "Bericht über die zoologische Station" (1879), 138.

30. Dohrn, "The Foundation" (1872), 437.

31. Coste, *Nidification* (1848).

32. His tasks were many, however, since he was also expected to take care of the fishermen, to deliver visiting scientists their research material, to supervise the public aquarium, and to keep records of the geography and life-cycles of the animals in the bay. Dohrn, *Erster Jahresbericht* (1876), 10.

33. Schmidtlein, "Beobachtungen über die Lebensweise" (1879).

34. Schmidtlein, *Leitfaden* (1880), 33.

35. Baedeker, *Italien: Handbuch* (1880), 82. For a satirized account of a guided tour by Schmidtlein: Buchholz [Stinde], *Buchholzens in Italien* (1884), 121-122.

36. Dohrn to Virchow, 6 Jan. 1884, Groeben and Wenig, *Anton Dohrn* (1992), 59. In 1890, the Naples aquarium would eventually inspire the French physiologist and chronophotographer Etienne-Jules Marey to take aquarium research in a different direction and set up studies into animal locomotion. In order to make films that registered the animal movements, however, Marey needed a particular set-up. He therefore created a special aquarium at his Naples villa that he filled with creatures he borrowed from the Stazione, rather than working in Dohrn's public aquarium. Marey, "Locomotion" (1890); Braun, *Picturing Time* (1992), 158.

37. Dohrn to Virchow, 15 Jan. 1879, Groeben and Wenig, *Anton Dohrn* (1992), 44.

38. Dohrn to du Bois-Reymond, 27 July 1879, Groeben et al., *Emil du Bois-Reymond* (1985), 158.

39. Dohrn, "Bericht über die zoologische Station" (1881), 500-501.

40. Dohrn to du Bois-Reymond, 27 July 1879, Groeben et al., *Emil du Bois-Reymond* (1985), 158.

41. MacLeod, "La station zoologique" (1882), 19.

42. Van den Broeck, "Une visite" (1882), 9-10.

43. In 1882, for example, he would take prince Heinrich of Prussia on an expedition. Groeben, *Charles Darwin* (1982),106.

44. Ghiselin and Groeben, "A Bioeconomic Perspective" (2000), 280.

45. Dohrn in the introduction to Chun, *Die Ctenophoren* (1880), VI.

46. Dohrn, "Bericht über die zoologische Station" (1886), 100.

47. Dohrn, "Bericht über die zoologische Station" (1893), 635.

48. Dohrn, "Bericht über die zoologische Station" (1881), 489.

49. Dohrn, "Bericht über die zoologische Station" (1886), 104-112.

50. Dohrn, "Bericht über die zoologische Station" (1886), 112-113.

51. Dohrn to Virchow, 6 Jan. 1884, Groeben and Wenig, *Anton Dohrn* (1992), 58-59.

52. Dohrn, "Bericht über die zoologische Station" (1893), 63.

53. Gieryn, "Three Truth-Spots" (2002), 128.

54. Dohrn, "Bericht über die zoologische Station" (1879), 137-139 and 150-152; "Some Unwritten History" (1897).

55. The American physician Simon Flexner recalled this statement of Dohrn in a letter to politician Christian Herter, from 21 Nov. 1903, quoted in: Groeben, *Charles Darwin* (1982), 23.

56. Dohrn explicitly stated that he had no pretension of "being the master" over his visitors. Boveri, *Anton Dohrn* (1910), 25.

57. Dohrn to du Bois-Reymond, 6 Jan. 1882, Groeben et al., *Emil du Bois-Reymond* (1985), 225.

58. Dohrn, "Bericht über die zoologische Station" (1886), 101.

59. I borrow the term from Lynn Nyhart: Nyhart, *Biology Takes Form* (1995), 205.

60. E.g., Dohrn, "The Foundation" (1872), 440.

61. A good example is the report the Belgian zoologist Julien Fraipont wrote to his master, Gustave Dewalque. Fraipont to Dewalque, 16 Jan. 1882, Ulg Dewalque Papers, 3319C. Jane Maienschein reported very similar reactions of Americans like Charles Otis Whitman. Maienschein, *Transforming Traditions* (1991), 102-103.

62. Whitman, "The Advantages" (1883), 92-96.

63. Dohrn, "Bericht über die Station" (1882), 593-594.

64. According to Dohrn the great majority would stay for less than six weeks.

65. Dohrn, *Erster Jahresbericht* (1876), 32-33.

66. Nunn Whitman, "The Zoological Station" (1886), 796.

67. Dohrn, "Bericht über die zoologische Station" (1886), 147-148.

68. Yung, "Le laboratoire de zoologie" (1881), 234.

69. Dohrn, "The Foundation" (1872), 439.

70. Lo Bianco had entered the station in 1875 as a lab servant when only fourteen, and he was trained on the spot. For the quote: Herdman, *A History of Science* (1904), 122.

71. Herdman, "The Greatest Biological Station" (1901), 426.

72. For the role Naples station played as "a clearing house of organisms," see: Groeben, "The Stazione" (2002). For Lo Bianco's techniques: Lo Bianco, *The Methods* (1899).

73. Dohrn, "The Foundation" (1872), 438.

74. Traweek, *Beamtimes* (1988), 19-21; Ophir and Shapin, "The Place of Knowledge" (1991), 9-11; Gieryn, "Biotechnology's Private Parts" (1998).

75. On "placelessness": Kohler, *Landscapes and Labscapes* (2002), 7-8; Livingstone, *Putting Science in Its Place* (2003), 3.

CHAPTER THREE

1. Livingstone, *Putting Science in its Place* (2003), 86.

2. Billick and Price, "The Ecology of Place" (2010).

3. Bohn, *Alfred Giard* (1910), 9-12; Bouyssi, *Alfred Giard* (1999), 79-110.

4. Giard, "L'éducation du morphologiste" (1911 [1908]), 49-50.

5. Giard's pupils collected the articles of their master in a posthumously published volume under that title; they taught courses and edited book series labeled as "general biology" and they proclaimed the discipline to be the central subject of their *Bulletin scientifique*. See among others: Blaringhem et al., "Introduction" (1909); Bohn, "La biologie générale" (1912); Rabaud, "Le domaine et la méthode" (1919); Massart, *Éléments de biologie générale* (1921).

6. Giard, "Les tendances actuelles" (1905), 134.

7. This was different from what was happening in Germany, where *general biology* also became a popular term around the turn of the century, but where it was usually the cell, heredity, or evolution that were represented as the unifying topics of the biological enterprise. Good examples are: Hertwig, *Allgemeine Biologie* (1906); Chun and Johannsen, *Allgemeine Biologie* (1915). Despite the differences, both in France and Germany general biology was seen above all as an experimental enterprise. For Germany, see: Laubichler, "Allgemeine Biologie" (2006).

8. Geoffroy Saint-Hilaire, *Histoire naturelle* (1854), vol. 1, 20.

9. Giard, "L'évolution des sciences biologiques" (1905), 198-199.

10. Haeckel, *Generelle Morphologie* (1866), vol. 2, 186-187. The translation is taken from: Acot, *The European Origins of Scientific Ecology* (1998), 671.

11. Coleman, "Evolution into Ecology?" (1986); Bowler, *The Earth Encompassed* (1992), 365-373; Nyhart, *Modern Nature* (2009); Schwarz and Jax, "Early Ecology" (2011).

12. See: Lankester, "Zoology" (1888); Largent, "Bionomics" (1999). For the rise of (British) animal ecology: Crowfort, *Elton's Ecologists* (1992).

13. The controversial publicist Félix le Dantec, for example, was interested in more abstract biological philosophy, while Émile Guyenot and Giard's eventual successor Maurice Caullery largely concentrated on the study of heredity.

14. The *Bulletin Scientifique* was an existing journal, which would be "hijacked" by Giard. After having become its editor in 1878, Giard slowly turned the journal into the organ of his station toward the end of the 1880s. See: Debaz, *Les stations françaises* (2005), 210-214 and 260-269.

15. The untitled program was signed by Louis Blaringhem, Georges Bohn, Maurice Caullery, Charles Julin, Félix Le Dantec, Félix Mesnil, Paul Pelseneer, Charles Pérez, and Etienne Rabaud, and appeared in 1909 in *Bulletin Scientifique de la France et de la Belgique*, 43: i-ii.

16. In fact, Dahl himself made increasing use of the term *Oekologie* after 1910. Dahl, "Experimentell-Statistische Ethologie" (1898); Dahl, "Was ist ein Experiment" (1901); Dahl, "Die Zieler der vergleichende 'Ethologie'" (1902); Erich Wasmann, "Biologie oder Ethologie?" (1901). On Dahl: Nyhart, *Modern Nature* (2009), 312-314 and 340-343.

17. In seminal works, biologists like Victor E. Shelford and Charles C. Adams explicitly rejected the term *ethology*, preferring *animal ecology* instead. Both also kept a strong link with the plant geographers. Wheeler, "'Natural History'" (1902); Shelford, *Animal Communities* (1913), 32-33; Adams, *Guide* (1977 [1913]), 19.

18. Durant, "Innate Character" (1981), 163-164; Burkhardt, *Patterns of Behavior* (2005). Also in the overviews of the history of French ecology Giard's name usually is lacking. See, e.g., Matagne, "The French Tradition" (2011).

19. Examples of both phenomena are abundant in the work of the Giardists. On mimicry see, e.g., Giard, "Sur le mimétisme" (1911 [1896]). On convergent evolution, inter alia: Giard, "Convergence" (1894); Caullery, "Les yeux et l'adaptation" (1905); Rabaud, *Les phénomènes de convergence* (1925).

20. See inter alia, Cépède, "Recherches sur les infusoires" (1910), 365-385.

21. Giard, "La castration parasitaire" (1911 [1887]).

22. It has been indicated before that, in his predilection for neologisms, Giard closely resembled Ernst Haeckel, who coined dozens of terms. Caullery, "L'oeuvre scientifique" (1909), xxxiv.

23. Giard, "Sur le bourgeonnement" (1891).

24. Giard, "L'anhydrobiose" (1911 [1896]).

25. Boissay, "Le laboratoire" (1877), 129.

26. Wildfowler [Clements], *Shooting and Fishing Trips* (1876), vol. 1, 208-254 and idem, *Shooting, Yachting, and Sea-Fishing* (1877), vol. 1, 267-269.

27. Wildfowler [Clements], *Shooting and Fishing Trips* (1876), vol. 1, 238.

28. Giard, "Discours" (1913 [1874]), 7.

29. Shapin, "'The Mind in Its Own Place'" (1990); Gieryn, "Three Truth-Spots" (2002).

30. Knight, *Tonic Bitters* (1868), 142.

31. Triboudeau, *Monographie agricole* (1904), 121-202.

32. Boulay, *Révision* (1878), 6.

33. Michelant, *Boulogne-sur-Mer* (1880), 54.

34. Trenard and Hilaire, *Histoire de Lille* (1999), 33-113.

35. Shinn, "The French Science Faculty System" (1979), 331.

36. Bouyssi, *Alfred Giard* (1999), 123.

37. Giard, "Discours" (1913 [1874]), 7-8; Boissay, "Le laboratoire" (1877), 129.

38. Kohler, *All Creatures* (2006), 18.

39. Menuge-Wacrenier, *La côte d'Opale* (1986), 101-105.

40. Giard, "Le laboratoire du Portel" (1913 [1889]), 64-67; Giard, "Coup d'oeil sur la faune du Boullonais" (1913 [1899]), 91; Schmitt, "Les deux laboratoires" (2012).

41. Ménégaux, "Le laboratoire maritime" (1905), 478.

42. Bouyssi, *Alfred Giard* (1999), 195-197.

43. Massart, "Les naturalistes actuels" (1912), 951.

44. Bohn, *Alfred Giard* (1910), 18-21.

45. Ensch and Querton, "La station zoologique" (1897), 309.

46. Giard, "Un amphipode" (1911 [1908]), 489.

47. However, occasionally zoologists from other nationalities—such as Carl Vogt, Edwin Ray Lankester, Alexander Kovalevsky, and Augusto Gonzalez de Linares—would work there. Giard, "Le laboratoire du Portel," (1913 [1889]), 67-69.

48. In 1892, the Belgian botanist Jean Massart wrote to his master Léo Errera, professor at the University of Brussels, that there was room for only three people in the laboratory of Wimereux, but that often there were up to eight researchers at work. Massart to Errera, 6 Sept. 1892, Errera Papers, Archives of the Université Libre de Bruxelles.

49. Cartaz, "La station zoologique" (1900), 146.

50. Giard, "Discours" (1913 [1874]), 17.

51. Ensch and Querton, "La station zoologique" (1897), 308-309.

52. Bohn, *Alfred Giard* (1910), 10.

53. According to Maurice Caullery—Giard's pupil and successor at the Sorbonne—the late marriage of his master, followed by the loss of three children within a short space of time, very much detached his master from the cenacle in Wimereux. Telkes, *Caullery* (1993), 127.

54. Cf. Journal A. Note 1-311 (1905-1906), Archives of the Institut Pasteur, Fonds Casimir Cépède, Notebooks.

55. Giard, "Discours" (1913 [1874]), 63-74.

56. See, e.g., Ensch and Querton, "La station zoologique" (1897), 305-310; Bohn, *Alfred Giard* (1910), 15-17 and 38-39; Massart to Errera, 6 Sept. 1892, Errera Papers, Archives of the Université Libre de Bruxelles.

57. Kohler, *Landscapes and Labscapes* (2002), 43.

58. See among others, Bohn, *Des mécanismes* (1901); Acloque and Cépède, *Observations biologiques* (1910); Massart, "Les naturalistes actuels" (1912); Rabaud, *Éthologie* (1912); Pelseneer, *Essai d'éthologie* (1935). The spread of the ethological program will be dealt with in more depth in chapter 4.

59. Bohn, *Alfred Giard* (1910), 17.

60. Clark, *Prophets and Patrons* (1973), 66-121.

61. The family metaphor was not limited to Giard's cenacle, but widely popular in turn-of-the-century science. See: Van Bosstraeten, "Dogs and Coca-Cola" (2011).

62. See, e.g., Caullery, "L'oeuvre scientifique" (1909), xvi and xxxvi-xxxvii; Le Dantec, "Jules Bonnier" (1909), lxxvii; Bohn, *Alfred Giard* (1910), 15-25; Massart, "Les naturalistes actuels" (1912), 950.

63. Even, to an extent, literally so. Giard's wife, Annie Bond-Cooke, would (after her marriage in 1892) appear at the station and purportedly do translation work, while Cépède's wife, the science teacher Léonie Paccard, would actually perform scientific research there (but without leaving traces in the published literature). Many of the visiting scientists, such as Pelseneer, Caullery, and Massart, brought wife and children to Wimereux in summer so that beach holidays could be combined with ecological study. During these summer stays the families would regularly meet in an informal atmosphere and strengthen the ties of the circle. The firmness of the network is further indicated by several marriages: Giard's ex-pupil Félix Mesnil married Maurice Caullery's sister, Gustave Loisel, another old student of Giard married the sister-in-law of Adolphe Cligny; while the son of Giardist Charles Pérez married the sister of Paul Pelseneer. Bonnier was the best man at both Pelseneer's and Giard's wedding. Bouyssi, *Alfred Giard* (1999), 199 and 210-212.

64. Bétencourt, "Les hydraires" (1888); Giard, "Le laboratoire de Wimereux" (1913 [1888]), 64; Telkes, *Caullery* (1993), 131.

65. In addition to references to such excursions (e.g., Journal 8: note 1748, Journal 9: note 4046, Journal 10: note 5001), the notebooks contain evidence of Cépède's activity in the local *Société des sciences naturelles de Seine et Oise*. Such activity was not uncommon in Giard's network. Archives of the Institut Pasteur, Fonds Casimir Cépède, Notebooks.

66. On this shift in the British context: Alberti, "Amateurs and Professionals" (2001). On the role of learned societies for early ecology in France: Matagne, "The French Tradition" (2011).

67. Giard, "Le laboratoire de Wimereux" (1913 [1888]), 39.

68. Boissay, "Le laboratoire" (1877), 34.

69. Bohn, *Alfred Giard* (1910), 18.

70. Ensch and Querton, "La station zoologique" (1897), 306.

71. Lacaze-Duthiers to Léon Frédericq, 16 June 1876, Florkin and Théodoridès, *Henri de Lacaze-Duthiers* (1982), 8.

72. Boutan, *La photographie* (1900), 116-117. Similar imagery was used in the description of the station of Saint-Vaast-La-Hougue: Coupin, "Le laboratoire maritime" (1894), 344

73. Giard, "Particularités" (1911 [1878]), 512-513; Giard, "Le laboratoire de Wimereux" (1913 [1888]), 38-62.

74. Herzig, *Suffering for Science* (2005), 71.

75. The argument works equally well for Lacaze-Duthiers, who combined his bent for exploration and physically exhausting work with an ideal of self-sacrifice in science. See Lacaze-Duthiers, "Sur les laboratoires" (1898), 3-12.

76. Simon Schaffer already hinted at the importance of "the country-city opposition" in the history of laboratories. Schaffer, "Physics Laboratories"(1998), 149-150.

77. Ménégaux, "Le laboratoire maritime" (1905).

78. Contemporary postcards do show the hotel, however.

79. Giard, "L'éducation du morphologiste" (1911 [1908]), 46.

80. Giard, "L'évolution des sciences" (1905), 198-199; idem, "Sur l'habitat" (1913 [1904]), 252-253.

81. Giard, "La méthode expérimentale" (1911 [1896]), 200.

82. Giard, "Préface" (1911 [1896]), 3.

83. Telkes, *Caullery* (1993), 214.

84. Giard, "La méthode expérimentale" (1911 [1896]), 196.

85. Massart to Errera, 6 Sept. 1892, Errera Papers, Archives of the Université Libre de Bruxelles.

86. Giard, "L'éducation du morphologiste" (1911 [1908]), 53.

87. Giard, "Particularités" (1911 [1878]), 509-511.

88. Giard was usually rather vague about who he was actually aiming at with such attacks. If he dropped names, it was usually those of lesser gods in the profession—particularly Germans such as Reinhold Teuscher, Heinrich Simroth, or Wilhelm Lange. Giard, "Particularités" (1911 [1878]), 512-513.

89. Massart, "Les naturalistes actuels" (1912), 947 and 953.

90. E.g., Rabaud, "Le domaine et la méthode" (1919), 5-6.

91. Giard, "La méthode expérimentale" (1911 [1896]), 196.

92. Giard, "Les tendances actuelles" (1905), 166; idem, "L'éducation du morphologiste" (1911 [1908]), 53.

93. Giard, "Discours" (1913 [1874]), 6.

94. Pelseneer, "Alfred Giard" (1908), 223.

95. As well as personal observations, Cépède's notebooks contain letters relating to his fieldwork. For the references to contacts with nonprofessionals, see inter alia, Ernest Rousseau to Victor Willem, 9 Oct. 1906; J. Maillard to Casimir Cépède, 6 June 1910, "Journal 4. 1906-1907," note 901, and "13. Entomologie. Journal," note 9653, Cépède papers, Institut Pasteur.

96. Giard, "Sur le Peroderma" (1896 [1888]), 212.

97. Giard, "Nouvelles remarques" (1896 [1888]), 348; Giard, "Sur un convoi migrateur" (1896 [1889]), 219.

98. Giard, "La méthode expérimentale" (1911 [1896]), 196.

99. Giard, "Préface" (1911 [1896]), 11.

100. Giard, "La castration parasitaire" (1911 [1887]), 241-242 ; Giard, "Préface" (1911 [1896]), 27.

101. Daston and Lunbeck, "Introduction" (2011), 4.

102. Massart, "La biologie des inondations" (1922).

103. Giard and Bonnier, *Contributions* (1887).

104. Bohn, *Des mécanismes* (1901).

105. Cépède, "Recherches sur les infusoires " (1910).

106. Also Cuvier used the term *expérience*. The full paragraph (taken from the English translation of 1837) goes as follows: "The most effectual mode of observing is by comparison. This consists in successively studying the same bodies in the different positions in which Nature places them, or in comparison between their structures and the phenomena which they manifest. These various bodies are kinds of experiments ready prepared by Nature, who adds to or subtracts from each of them different parts, just as we might wish to do in our laboratories, and shows us herself in the results of such additions or retrenchments." Cuvier, *The Animal Kingdom* (1837), vol. 1, 3.

107. See among others, Outram, *Georges Cuvier* (1984), 62-63 and 137; idem, "New Spaces" (1995).

108. Giard, "De l'influence" (1911 [1889]).

109. Giard, "Sur la transformation" (1911 [1889]).

110. Massart, "La biologie et la végétation" (1893).

111. Kohler, *Landscapes and Labscapes* (2002), 212-213.

112. Telkes, *Caullery* (1993), 89.

113. Bohn, *La nouvelle psychologie* (1911), 37.

114. Cépède, *Convois migrateurs* (1912), 34-35.

115. For Bohn, see inter alia, Bohn, *La naissance de l'intelligence* (1909), 45-49 and 113-138. On Rabaud's tropism research, see: Etienne Rabaud, *Titres et travaux* (1935), 71-77.

116. Bohn, "A quoi peut-on reconnaître?" (1905).

117. Bohn, "La biologie générale" (1912), 362.

CHAPTER FOUR

1. Of the Giardists who performed substantial ecological research, Bohn, Rabaud, Cépède, Dollo, and Pelseneer saw neo-Lamarckian factors as at least partial causes of evolution. Only Massart defended neo-Darwinian ideas. On Giard's role in French neo-Lamarckism: Conry, *L'introduction du darwinisme* (1974), 240-241; Persell, *Neo-Lamarckism* (1999), 23-58; Loison, *Les notions de plasticité et hérédité* (2008); Loison, "French Roots" (2011). For the ideas of his pupils see inter alia, Bohn, "Le passé" (1908), 627; Cépède, "Les manuscrits de Lamarck" (1908); De Bont, *Darwins kleinkinderen* (2008), 206-213.

2. On Giard's political career and the convictions of some of his pupils, see: Bouyssi, *Alfred Giard* (1999), 215, 252-373, and 561. On the political standpoints of his Belgian followers: De Bont, *Darwins kleinkinderen* (2008), 219-224.

3. E.g., Giard, "Le principe de Lamarck" (1904 [1890]), 152.

4. On the importance of science for Third Republic ideology: Fox, *The Savant and the State* (2012), 227-273.

5. Giard and his pupils would play a key role in the rise of this cult in the following decades. They reintroduced Lamarck's ideas in historiography, they supported the attempts to erect his statue at the Muséum d'Histoire Naturelle (which succeeded in 1908), and they took the initiative of setting up a prize in his name at the Belgian Academy of Sciences (which was established in 1913). Giard, "L'évolution des sciences biologiques" (1905), 201; Bouyssi, *Alfred Giard* (1999), 407-408 and 631-635; De Bont, *Darwins kleinkinderen* (2008), 228-229.

6. Donnat and the Parisian chair for evolution of organized beings are extensively discussed in: Bouyssi, *Alfred Giard* (1999), 408-423.

7. De Bont, *Darwins kleinkinderen* (2008), 221.

8. For an overview of the positivist climate in Belgium and the University of Brussels in particular: Daled, *Spiritualisme et matérialisme* (1998); Wils, *De Omweg van de wetenschap* (2005).

9. De Bont, *Darwins kleinkinderen* (2008), 193-194.

10. E.g., Giard "Les facteurs de l'évolution" (1904 [1889]), 120.

11. On the "solidarist" use of Lamarckism in the French Third Republic: La Vergata, "Lamarckisme et solidarité" (1996); Bernardini, *Le darwinisme social en France* (1997), 139-148; Farber, "French Evolutionary Ethics" (1999); Gissis, "Late Nineteenth Century Lamarckism" (2002); Thomas, *Rethinking the History* (2003), 111-127.

12. Pelseneer, "La morale de la science" (1902-1903).

13. "Discussion de l'interpellation de M. Giard" (1884), 651.

14. Sherard, *Émile Zola* (1893), 201.

15. Zola, *Germinal* (1886); idem, *Carnets d'enquêtes* (1987).

16. See: Béguet, "La vulgarization" (1994), 13-16.

17. See Giard's Preface in Loeb, *La dynamique* (1908), VII.

18. E.g., Bohn, *La naissance de l'intelligence* (1909).

19. See e.g., Wéry, *Excursions scientifiques: Sur le littoral belge* (1908), 4 and 7-10; Schouteden-Wéry, *Excursions scientifiques. II. En Brabant* (1913), 41. On Massart's role for the nature conservation movement: De Bont and Heynickx, "Landscapes of Nostalgia" (2012).

20. "2. Journal B," note 695, Cépède Papers, Institut Pasteur. Kuckuck for his part was the first curator for botany at the marine station in Helgoland at the Baltic Sea.

21. Capus, *Guide du naturaliste* (1903), x.

22. Giard, *Exposé des titres et travaux* (1896), 7.

23. Raveret-Wattel, "La station aquicole" (1889); Cligny, "Sur l'éthologie du hareng" (1904); Kofoid, *The Biological Stations* (1910), 123-126; Bouyssi, *Alfred Giard* (1998), 215-216.

24. Giard, "Un enemi peu connu" (1896 [1877]).

25. Giard, "L'éducation du morphologiste" (1911 [1908]), 51.

26. Giard, "Considérations sur les insectes" (1896 [1876]).

27. It was probably Metchnikoff who brought Krassilstchik in contact with Giard. In 1888 the latter would publish a translated article by Krassilstchik in his *Bulletin biologique*. Krassilstchik, "La production industrielle" (1888). On the tradition of biologically dealing with insect pests, see: Lord, "From Metchnikoff to Monsanto" (2005).

28. See: Giard, "Deux ennemis" (1896 [1881]); "Sur le *Silpha Opaca*" (1896 [1888]); Giard, "Les saumons de la Canche (1896 [1888]); Giard and Roussin, "Comité consultatif" (1896 [1889]); Giard and Roché, "Rapport adressé au Ministère" (1896 [1895]).

29. Jaynes, "The Historical Origins" (1969), 602-603.

30. Janet, "Société internationale" (1900-1901); "Allocution de M. Pierre Janet" (1900-1901), 133-139; *Institut général psychologique* (n.d. [1909]). On the relationship between spiritualism and psychology at the Institute, see: Marmin, "Métapsychique et psychologie" (2001), 157-158. On Third Republic ideology, see Brower: *Unruly Spirits* (2010), 57-59.

31. Bohn, "De la recherche des abris" (1903); Rabaud, "Observations sur les manifestations mentales" (1904); Giard, "Les origines de l'amour" (1911 [1905]).

32. Bohn, *Attractions & oscillations* (1905).

33. Giard, "Les origines de l'amour" (1911 [1905]).

34. Bohn, *La naissance de l'intelligence* (1909), 69-79 and 139-180; Bohn, *La nouvelle psychologie* (1911), 27-40.

35. Bohn, "Les progrès récents" (1911), 396.

36. See inter alia, Washborn, "La nouvelle pyschologie" (1912); Haggerty, "La nouvelle pyschologie" (1912).

37. In her work on French animal behavior studies Marion Thomas connects Bohn's marginalization to his conflict with the (more successful) psychologist Henri Piéron. Piéron also followed courses with Giard, but did not stay so close to his ecological program. Thomas, *Rethinking the History* (2003), 74-103.

38. Massart's organicism was obviously not an exceptional stance in the late-nineteenth-century Francophone world. In left-leaning, republican circles it was quite common. For this organicism, the writings of Herbert Spencer in particular proved an important source of inspiration. See inter alia, Beck, "The Diffusion of Spencerism" (2004).

39. Massart and Vandervelde, "Parasitisme organique" (1893). The article also appeared in the left-wing *La Société Nouvelle* and was later published as a book in English, Italian, and French.

40. Demoor et al., *L'évolution régressive* (1897).

41. Frost, *The Functional Sociology* (1960), 99-100; Crombois, *L'univers de la sociologie* (1994), 33-38.

42. Waxweiler, *Esquisse d'une sociologie* (1906), 62, 75, and 263.

43. Waxweiler, *Esquisse d'une sociologie* (1906), 106, 263, and 270; idem, "La vie" (1974 [1906]); idem, "Avant-propos" (1910).

44. Waxweiler and his pupils were influenced among other things by the innovative anthropological work of Franz Boas, whose criticism of racial hierarchies they would take up.

45. As an example of biological language hiding old-style human science, see: Warnotte, "Le conflit des adaptations dans l'évolution sociale" (1911). A good illustration of deterministic

physical anthropology somewhat uncomfortably brought under the umbrella of "social ecology" is: Houzé, *L'Aryen* (1906).

46. Crombois, *L'univers de la sociologie* (1994), 117-122.

47. On Dollo: Brien, "Notice sur Louis Dollo" (1951); Gould, "Dollo" (1970).

48. In 1909, he also unsuccessfully tried to have Dollo appointed director of the Brussels natural history museum. Cf. draft regarding the directorship of the natural history museum, Dorlodot papers, Archives de l'Université Catholique de Louvain.

49. To develop a paleontological application of "ecological" thinking, Dollo turned not so much to the "Giardists" as to the writings of the Russian paleontologist Vladimir O. Kovalevskii. Particularly his work on ungulates from the 1870s, which studied the adaptive advantages of the evolutionary reduction in the number of toes, was hailed as an example. Dollo, "La paléontologie éthologique" (1909), 377 and 383-386. On Kovalevskii: Todes, "V. O. Kovalevskii" (1978).

50. Sues, "European Dinosaur Hunters" (1997), 22.

51. This mode of representation was initiated by Robert Owen. See: Desmond, "Designing the Dinosaur" (1979).

52. Dollo, "Troisième note sur les dinosauriens de Bernissart" (1883), 118-119.

53. Dollo, "Les lois de l'évolution" (1893), 164-165.

54. Louis Dollo, "Les peupliers," Cours de Mr. L. Dollo Physiologie 1897-1898, Archives de l'Université Libre de Bruxelles, 1Q 3638.

55. Especially after the First World War, Dollo became rather isolated. Among other things, he was reproached for not having signed a protest action of the Academy against the deportation of Belgians under the German occupation. Léon Fréдeriсq to Dollo, Feb. 8th 1919, and undated draft of Dollo, RBINS, Dollo Papers.

56. Abel, "Die Festgabe" (1928); idem, "Louis Dollo" (1928); Gould, "Dollo" (1970), 189-212.

57. Loisel, *Exposé des titres* (1905); Telkes, *Caullery* (1993), 130 and 327.

58. Loisel, *Projets et études* (1907); Bouyssi, *Alfred Giard* (1998), 496 and 643-645.

59. On Hagenbeck's zoo: Rothfels, *Savages and Beasts* (2002).

60. Loisel, "Rapport sur une mission scientifique" (1908), 367-388.

61. Loisel, "Rapport sur une mission scientifique" (1908), 382-384.

62. Telkes, *Caullery* (1993), 130.

63. Coulon, *Les ennemis* (1925), 26-29, cf. Bouyssi, *Alfred Giard* (1998), 44-55.

CHAPTER FIVE

1. De Guerne, "Le laboratoire" (1892), 149.

2. E.g., McIntosch, *The Background of Ecology* (1985), 57-61 and 120-127; Leps, "Ökologie" (1998).

3. Zacharias, "Über die systematische Durchforschung" (1905), 3.

4. The unfolding of the plans can be followed in Zacharias's elaborate letters to Darwin: Zacharias to Darwin, 3 Jan. 1875, 19 August 1875, 7 Jan. 1877, 23 Feb. 1877, CUL DAR 184:1-5.

5. On the *Kosmos-Darwiniana* case: Daum, "Naturwissenchaftlicher Journalismus" (1995); Wetzel and Nöthlich "Vom 'Homo literatus' zu 'Self-made man'" (2006), 466-469. The quotation comes from a letter from Haeckel to Zacharias of 18 May 1875, published in: Nöthlich et al. "'Ich acquirirte das Schwein'" (2006), 199.

6. Haeckel to Friedrich von Hellwald, 18 Dec. 1876, published in Di Gregorio, *From Here to Eternity* (2005), 396.

7. The citations come respectively from a letter from Krause to Haeckel (10 Dec. 1876) and one from Haeckel to Zacharias (17 Jan. 1877), cited in Wetzel and Nöthlich, "Vom 'Homo literatus' zu 'Self-made man'" (2006), 469.

8. A. Thienemann, "Otto Zacharias †" (1917); Nöthlich et al. "'Ich acquirirte das Schwein'" (2006).

9. A complete bibliography can be found in: A. Thienemann, "Otto Zacharias †" (1917), xiv-xxiv.

10. Zacharias to Haeckel, 14 Jan. 1885, published in: Nöthlich et al. "'Ich acquirirte das Schwein'" (2006), 241.

11. This despite the fact that Zacharias's personal relation with Dohrn had turned sour after the latter had bluntly refused to cooperate with any Darwinian popularizing. Zacharias, "Vorwort" (1897), v. See also: Zacharias, "Das 25-jährige Jubiläum" (1897), 485; Zacharias to Haeckel, 11 June 1875, published in: Nöthlich et al. "'Ich acquirirte das Schwein'" (2006), 201.

12. Zacharias, "Ueber die Errichting" (1889); Zacharias, "Die biologische Station" (1891); Zacharias, "Die biologische Station" (1892), 38; Zacharias, "Vorwort" (1896), vi-vii; Zacharias, "F. A. Krupp als Freund" (1903).

13. Du Bois-Reymond to Zacharias, 24 July1893, Universitätsbibliothek Kiel, Nachlass Zacharias. On Zacharias's Berlin contacts, see: Daum, *Wissenschaftspopularisierung* (1998), 401-402.

14. Zacharias, "Hydrobiologische Aphorismen" (1894), 148.

15. In 1888, for example, the German Fishery Union donated 10,000 Deutschmark (the equivalent of about 180,000 dollars in 2013) to support the oceanographic Plankton-Expedition—which was carried out with a steamship significantly named the "National." The same Union made crucial contributions to the founding of the Helgoland marine station four years later. Heincke, "Die Biologische Anstalt" (1893); Mills, *Biological Oceanography* (1989), 23-27.

16. Zacharias, "Die biologische Station" (1891), 108-109.

17. Zacharias, "Über die lacustrisch-biologische Station" (1889), 600; idem, "Über die eventuelle Nützlichkeit" (1906), 294.

18. This isolation was obviously a rhetorical construct. For one, the river Schwentine passes through the Großer Plöner See.

19. Zacharias, "Über die lacustrisch-biologische Station" (1889); Zacharias, "Über die systematische Durchforschung" (1905), 2; Zacharias, "Vorwort" (1905), viii.

20. Baedeker, *Mittel- und Nord-Deutschland* (1885), 90-92; Zacharias, "Über die wissenschaftliche Aufgaben" (1891), 329; Zacharias, "Die biologische Station" (1892); Walter, "Biologie" (1894), 197; Zacharias, "Ferienkurse" (1908), 271-272. On the rise of German lakeside resorts and steamship-tourism: Blackbourne, *The Conquest of Nature* (2006), 159-162.

21. Zacharias, "Die biologische Station" (1892), 39. For the history of the Holstein railway network: Lange, "Modernisierung" (1996), 348-352.

22. See: Brandt, *Geschichte Schleswig-Holsteins* (1981), 294-301; Lorenzen-Schmidt, "Bevölkerungsentwicklung" (1996); idem, "Neuorientierung" (1996). A lively description of Holstein around 1880 (and further references to the existing literature on the history of the *Heimat* concept) can be found in: Nyhart, *Modern Nature* (2009), 11-15.

23. Zacharias, "Summarischer Bericht" (1888), 186; idem, "Hydrobiologische Aphorismen" (1894), 148; idem, "Vorwort" (1895), vii.

24. Zacharias, "Ueber die Errichtung" (1889), 23; idem, "Über die wissenschaftlichen Aufgaben" (1891), 315.

25. Rossmässler, *Das Süsswasser-Aquarium* (1857). On Rossmässler: Daum, "Science" (2002); Brunner, *The Ocean at Home* (2003), 59-63.

26. Zacharias, "Über die systematische Durchforschung" (1905), 10.

27. The most important of these were probably the Norwegian Michael Sars, the Swede Wilhelm Lilljeborg, and the Dane Erasmus P. Müller. None of these men focused on freshwater research alone, however. Mostly it was just a side-project of marine zoology.

28. On the idea of the lake as a zoological "wasteland," see: Schwarz, *Wasserwüste* (2003). On the role of Forel: Egerton, "The Scientific Contributions" (1962); Steleanu, *Geschichte der Limnologie* (1989), 100-110.

29. See Forel to Zacharias, 17 Feb. 1889, 21 Feb. 1890, and 2 March 1890, Universitätsbibliothek Kiel, Nachlass Zacharias. Zacharias, "Über die systematische Durchforschung" (1905), 8–9.

30. Hensen, "Über die Bestimmung des Planktons" (1887).

31. Zacharias, "Ueber das Einsammeln" (1889), 325.

32. Zacharias, "Das Plankton" (1905), 280.

33. See among others: Krauße, "Haeckel" (1995); Köckerbeck, *Die Schönheit* (1997); Bayertz, "Biology and Beauty" (1999); Steigerwald, "The Cultural Enframing" (2000); Steigerwald, "Goethe's Morphology" (2002); Stephenson, "'Binary Synthesis'" (2005); Di Gregorio, *From Here to Eternity* (2005), 516–518.

34. Zacharias, "Vorschlag zur Gründung" (1888), 19 and 24.

35. Zacharias, "Das Plankton" (1905), 274–276.

36. Zacharias, "Vorwort" (1895), v.

37. E.g., Zacharias, "Ueber das Einsammeln" (1889), 306; Zacharias, "Die moderne Hydrobiologie" (1905), 85.

38. Forbes, "The Lake as a Microcosm" (1887). More on Forbes and his microcosm concept: Schneider, "Local Knowledge" (2000); Schwarz, *Frühe Ökologie* (2000), 267–270.

39. Zacharias, Die biologische Station (1904), 3.

40. Zacharias, "Ueber das Einsammeln" (1889), 306.

41. Zacharias, "Das Plankton" (1905), 266.

42. Rossmässler, *Das Süsswasser-Aquarium* (1857), 8–9 and 52.

43. Junge, *Der Dorteich* (1885). About "the spread of the Village Pond gospel": Nyhart, *Modern Nature* (2009), 173–197.

44. Zacharias, "Ueber das Einsammeln" (1889).

45. Zacharias, "Das Plankton" (1905), 266.

46. Zacharias, "Über die eventuelle Nützlichkeit" (1906), 247.

47. By the time Zacharias took it up, the topos of "nature as a laboratory" was obviously very old already and it had definitely not been used only by life scientists. It proved particularly useful, however, to serve the cause of the "new naturalists" in the late nineteenth century. For the use of the metaphor among German plant ecologists: Cittadino, *Nature as the Laboratory* (1990).

48. Zacharias, "Summarischer Bericht" (1888), 188.

49. Zacharias, "Über die lacustrisch-biologische Station" (1889), 601.

50. Zacharias, "Ueber das Einsammeln" (1889), 306; idem, "Die Aufgaben" (1889), 115; Walter, "Biologie" (1894), 141; Zacharias, "Über die systematische Durchforschung" (1905), 13; idem, "Über die eventuelle Nützlichkeit" (1906), 245–246.

51. Zacharias to Haeckel, 4 April 1898, published in: Nöthlich et al. "'Ich acquirirte das Schwein'" (2006), 243. As Lynn Nyhart has indicated, even Haeckel was not entirely happy with the stress on "section cutting" among his students, and the lack of attention for "the entire animal" and "its mode of life." Nyhart, *Biology Takes Form* (1995), 203.

52. Zacharias, "Die moderne Hydrobiologie" (1905), 85.

53. Zacharias, "Über die eventuelle Nützlichkeit" (1906), 267.

54. Zacharias, "Über die eventuelle Nützlichkeit" (1906), 267. The involvement of untrained fishermen was not uncontroversial. See: Apstein, "Antwort" (1893), 243.

55. Zacharias, "Ueber die Errichtung" (1889), 23–25; idem, "Die Aufgaben" (1889), 116; idem, "Vorwort" (1898), viii–ix.

56. This, he stressed, was particularly on account of Eilhard Schulze (Berlin), Carl Chun (Leipzig), and Viktor Hensen (Kiel). Zacharias, "Das Plankton" (1905), 265–266.

57. Meyer and Möbius, *Die Fauna* (1865 and 1872). Möbius's pre-Kiel career is extensively discussed in: Nyhart, *Modern Nature* (2009), 125–145.

58. I deliberately used a more literal translation than the (still often used) English translation of H. J. Rice, which appeared in 1883. For the German original, see: Möbius, *Die Auster* (1877), 76; for Rice's translation: Möbius, "The Oyster" (1883).

59. On Möbius's Biocönose concept, see, next to Nyhart's work: Potthast, "Historische und ökologietheoretische Perspektiven" (2006); Leps, "Karl August Möbius" (2006).

60. Hensen and his school have been extensively researched in: Mills, *Biological Oceanography* (1989), 1-186; idem, "The Ocean Regarded as a Pasture" (1990).

61. The Silesian report also played a central role in the awakening of Virchow's liberal ideology. See, e.g., Weindling, "Was Social Medicine Revolutionary?" (1984), 14-16; McNeely, *"Medicine on a Grand Scale"* (2002), 14-19.

62. Up to today the 1879 famine remains little researched. For a detailed history of industrial labor in Upper Silesia: Schofer, *The Formation* (1975). For contemporary commentary, e.g.: Kräcker, *Etwas mehr Licht* (1880); Sucker, *Der Nothstand* (1880); Wocke, *Regierung* (1880).

63. Zacharias, *Die Bevölkerungs-Frage* (1880).

64. Zacharias, "Vorschlag zur Gründung" (1888), 21.

65. Both obviously shared a background in popular biology. Möbius would stress this himself in the answer of a congratulatory letter he received from Zacharias for his seventieth birthday. Zacharias to Möbius, 15 August 1886, Zoologisches Museum Kiel, Nachlass Möbius; Möbius to Zacharias, 29 January 1895, Universitätsbibliothek Kiel, Nachlass Zacharias.

66. Möbius to Brandt, 21 February 1888, and Zacharias to Brandt, 27 April and 19 August 1888, Zoologisches Museum Kiel, Nachlass Brandt. The letters from Brandt to Zacharias have not been preserved, but summaries have been recorded in the "Briefbuch" of the Zoologisches Museum, Nachlass Brandt.

67. Zacharias, "Vorwort" (1891), v.

68. Apstein, "Die quantitative Bestimmung" (1891); Zacharias, "Die Fauna" (1892).

69. Zacharias, "Vorläufiger Bericht" (1892), 578.

70. Apstein's article led to a reply of Zacharias in *Die Heimat*, which was followed by new accusations by Apstein. Apstein, "Veröffentlichungen" (1893); Zacharias, "Entgegnung" (1893); Apstein, "Antwort" (1893).

71. Zacharias to Brandt, 16 July 1893, and Brandt to Zacharias (draft), 18 July 1893, Zoologisches Museum, Nachlass Brandt.

72. *Acht fachmännische Gutachten* (1894).

73. Zacharias, "Ausweis" (1898), 215-216.

74. Lemmermann, Walter, Reichert, and Voigt left no traces in the standard biographical reference works. The careers of these men therefore can be reconstructed only through their publication record and short references other authors make to them. Some complementary information on Lemmermann and Voigt can be found in: Voigt, *Die Rotatorien* (1904), 194; Zacharias, "† Ernst Lemmerman" (1917).

75. Among others, Zacharias, "Quantitative Untersuchungen" (1896).

76. Zacharias, *Das Süsswasser-Plankton* (1907), 84-91.

77. Zacharias, "Hydrobiologische Aphorismen" (1894), 149-150. See also: Zacharias, "Vorwort" (1895), vi.

78. Zacharias, "Beobachtungen" (1894); idem, "Untersuchungen" (1898); "Über die jahreszeitliche Variation" (1903).

79. When the first five-year state subsidy was threatened not to be prolonged, several regional fishery societies worked together to set up a mass petition addressed to Reichskanzler Bismarck, while a committee of Hamburg merchants started a fund-raising. See the brochure *Aufruf an sämtliche Fischereivereine Deutschlands in Betreff der Biologische Station zu Plön* and the newspaper cuttings for the year 1897 in the archives of the Max-Planck-Insitut für Evolutionsbiologie, Plön.

80. Zacharias, "Über die eventuelle Nützlichkeit" (1906), 265.

81. Zacharias, "Vorwort" (1896), ix–x; idem, "Über die systematische Durchforschung" (1905), 33–34.

82. Zacharias, "Über die eventuelle Nützlichkeit" (1906), 252.

83. See among others, Ringer, *The Decline* (1969), 253–366; Goldman, *Politics* (1992), 25–50; Ringer, *"Bildung"* (2000).

84. Zacharias, "Das Plankton als Gegenstand" (1905), 251.

85. Lampert, *Das Leben* (1899); Zacharias, *Das Süsswasser-Plankton* (1907); Francé, *Der Bildungswert* (1907); Seligo, *Tiere und Pflanzen* (1908).

86. Zacharias, "Das Plankton als Gegenstand" (1905), 272 and 291–293; idem, "Die staatliche Sanktion" (1908), 235; idem, "Biologische Schülerübungen" (1908), 273.

87. Zacharias, "Ferienkurse" (1908); idem, "Der Ferienkurse-Pavillon" (1910).

88. Rieper, "Über die Ferienkurse" (1911). Other complaints were published in *Monatsheften für den naturwissenschaftlichen Unterricht, Süddeutschen Schulblättern,* and *Preussischen Lehrerzeitung.*

89. Zacharias, "Zur Entgegnung" (1911); idem, "Über den speziellen Zweck" (1911); Rieper, "Antwort" (1911); Zacharias, "Ein letztes Wort" (1911); Rieper, "Nochmals" (1911).

90. See, e.g., Seligo, *Hydrobiologische Untersuchungen* (1890), 3 and 44; Lampert, *Das Leben* (1899), 25; Francé, *Das Leben,* vol. 3 (1906), ix and 517.

91. Among others, Ward, "The Fresh-Water Biological Stations" (1899), 504; Rousseau, "Avant-Propos" (1906), x.

92. Zacharias, "Über chromatophile Körperchen" (1912); idem, "Zur Cytologie" (1912); idem, "Über den feineren Bau" (1913).

93. A. Thienemann, *Erinnerungen* (1959), 92.

94. See: Vom Brocke, "Die Kaiser-Wilhelm-/Max-Planck-Gesellschaft" (1996).

95. The most detailed source on Thienemann's life remains his autobiography: A. Thienemann, *Erinnerungen* (1959). See also: Schneller, *Das Werk* (1993); Egerton, "Thienemann" (2008); Schwarz and Jax, "Early Ecology" (2011), 235–242.

96. A. Thienemann, *Erinnerungen* (1959), 78–261.

97. Early key publications are: A. Thienemann, "Seetypen" (1921); idem, "Der *Produktionsbegriff*" (1931).

98. See, e.g., the back cover of Schneller's *Das Werk* (1993). Or: Leps, "Ökologie" (1998), 607 and 610.

99. A. Thienemann, "Otto Zacharias" (1917), 12; idem, "Zehn Jahre" (1927), 58.

100. A. Thienemann, "Die wissenschaftlichen Aufgaben" (1917), 625.

101. E.g., A. Thienemann, "Lebensgemeinschaft" (1918).

102. Among others, Walter, *Die Fischerei* (1903); Zacharias, *Das Süsswasser-Plankton* (1907), 84–85.

CHAPTER SIX

1. The station of Plön was, among others, described as an explicit example in: J. Thienemann, *Die Vogelwarte* (1910), 6.

2. See: Möller, "Notizen" (2004), 154.

3. See among others: Haffer et al. *Erwin Stresemann* (2000); Kruuk, *Niko's Nature* (2003); Burkhardt, *Patterns of Behavior* (2005); Haffer, *Ornithology* (2008).

4. Although no central figure in the historiography of science, some commemoration literature exists concerning Johannes Thienemann. See for instance: Schüz, "Johannes Thienemann" (1938); Berthold and Schlenker, "Johannes Thienemann" (1995); Möller, "Notizen" (2004).

5. Their first common ancestor is the early-eighteenth-century pastor Johannes Gottfried Thienemann, who was Johannes's great-great-grandfather and August's great-great-great-grandfather. Cf.: http://www.thienemann-archive.org/index.php, consulted 17 August 2010.

6. Quetelet, *Letters* (1846), 296.

7. "Notice sur les hirondelles" (1838). On Quetelet's ambitious project of mapping all periodical phenomena: Demarée and Chuine, "A Concise History" (2006).

8. Stresemann, *Die Entwicklung* (1951), 338–341; Allen, *The Naturalist in Britain* (1994), 198–200.

9. Gätke, *Die Vogelwarte* (1891); Hünemörder, "Ornithology" (1995).

10. Floericke, "Die Gründung" (1894). On Floericke see: Franke, *Dr. Curt Floericke* (2009).

11. In 1902 already, Floericke published a pamphlet that attacked the Vogelwarte. In it, he showed himself particularly angry that his early work had not been referenced by Thienemann, but was nonetheless heavily used. Furthermore Floericke stressed that Thienemann's ornithological publications did not surpass the level of those of the average schoolmaster and that, despite phrases about bird protection, he killed birds "en masse." In his view the Vogelwarte was nothing but "a sinecure for a theologian gone to the dogs." Floericke, *Kritik* (1902).

12. One gets an impression of the group's composition in Haffer, "Gruppenbilder" (2003).

13. Möller, "Notizen" (2004), 150–154. On Blasius and Rörig—and indeed most Central-European ornithologists—good biographical entries can be found in: Gebhart, *Die Ornithologen* (2006).

14. McElligot, *The German Urban Experience* (2001), 1.

15. For a good overview of the history of East Prussia: Kossert, *Ostpreussen* (2007).

16. Bezzenberger, *Die Kurische Nehrung* (1889), 175–285; J. Thienemann, *Rossitten* (1930), 72–73. According to Tinbergen fifty people still spoke the old Kurish language when he visited the place in 1925. He also indicated German was slowly taking over from the Baltic languages. Tinbergen to his parents, 30 Sept. 1925, Bodleian Library, Ms. Eng. c. 3125, File 1. 14.

17. Schmidt and Blohm, *Die Landwirtschaft* (1978), 500; Blackbourn, *History of Germany* (2003), 152; Wagner, *Bauern* (2005); Kossert, *Ostpreussen* (2007), 138–193.

18. J. Thienemann, "I. Jahresbericht" (1902), 149.

19. The report of Lindner from 1899 is quoted in Möller, "Notizen" (2004), 154.

20. Floericke, "Die Gründung" (1894), 233–235; J. Thienemann, "IX. Jahresbericht" (1910), 534–535.

21. J. Thienemann, *Die Vogelwarte* (1910), 6.

22. On this theme, see: Shapin, "'The Mind in Its Own Place'" (1990); Gieryn, "Three Truth-Spots" (2002).

23. J. Thienemann, *Rossitten* (1930), 14.

24. J. Thienemann, "I. Jahresbericht" (1902), 143–145.

25. J. Thienemann, "I. Jahresbericht" (1902), 207; idem, "X. Jahresbericht" (1911), 623.

26. Berlepsch, *Der gesamte Vogelschutz* (1899). See also: Schmoll, "Indication" (2005).

27. E.g., Rörig, *Magenuntersuchungen* (1900). See also: Jansen, *"Schädlinge"* (2003), 165–170.

28. Möller, "Notizen" (2004), 156.

29. Some of the results of these analyses are discussed in: Schalow, *Verhandlungen* (1911), 103.

30. Schmoll, "Indication" (2005). On similar developments in the United States: Worster, *Nature's Economy* (1985), 258–290; Barrow, *A Passion* (1998), 146–148.

31. J. Thienemann, *Das Leben* (1939), 15.

32. J. Thienemann, "I. Jahresbericht" (1902), 204–208; idem, "II. Jahresbericht" (1903), 225–228.

33. J. Thienemann, "II. Jahresbericht" (1903), 163–164.

34. J. Thienemann, "Wie die alten sungen" (1936), 174.

35. Herman, "Kurze Übersicht" (1911), 39.

36. On Heincke's biometric herring research: Jansen, "Den Heringen" (2002).

37. In the United States one generally uses the term *bird banding*, whereas in the United Kingdom the term *bird ringing* is used for the same practice.

38. J. Thienemann, "XX. Jahresbericht" (1922), 61.

39. J. Thienemann to Herman Schalow, 21 Nov. 1903, Nachl. Schalow, Staatsbibliothek Berlin.

40. J. Thienemann, "IV. Jahresbericht" (1905), 397-399; idem, "V. Jahresbericht" (1906), 468.

41. J. Thienemann, "III. Jahresbericht" (1904), 281.

42. J. Thienemann, Rossitten (1930), 149.

43. J. Thienemann, "V. Jahresbericht" (1906), 462.

44. J. Thienemann, "VI. Jahresbericht" (1907), 531; idem, Rossitten (1930), 218-219.

45. Floericke, Jahrbuch (1908), 40-41; Löns, Mümmelman (1909), 25; idem, "Magenuntersuchung" (1910), 440-441. On Floericke and Löns as science popularizers: Daum, Wissenschaftspopularisierung (1998), 325, 375, 396, and 420.

46. See among others, Doughty, Feather Fashions (1975); Schmoll, "Indication" 2005.

47. J. Thienemann, "VII. Jahresbericht" (1908), 406.

48. J. Thienemann, "XI. Jahresbericht" (1913), 2.

49. See, e.g., Lowe, "Amateurs" (1976), 517-535.

50. Nyhart, Modern Nature (2009), 243-244.

51. J. Thienemann, "V. Jahresbericht" (1906), 431.

52. These included, among others, Johan Axel Palmén (Finland), Eduard van Oort (the Netherlands), Robert Poncy (Switzerland), Jakob Schenk (Hungary), and Dimitri von Kaygodoroff (Russia).

53. J. Thieneman, "X. Jahresbericht" (1912), 133-243.

54. See for instance, J. Thienemann, "IX. Jahresbericht" (1910), 620-624; idem, "X. Jahresbericht" (1912), 144-147.

55. Kossert, Ostpreussen (2007), 217-231.

56. J. Thienemann, "XIX. Jahresbericht" (1921), 13; idem, "XX. Jahresbericht" (1922), 62.

57. On the role of the Heimat concept among German conservationists: Williams, "The Chords" (2006), 339-384.

58. J. Thienemann, "XI. Jahresbericht" (1913), 8; idem, "XX. Jahresbericht" (1922), 65-66; idem, "XXI. Jahresbericht" (1923), 145-146; idem, Im Lande (1933), 63. See also: Von Lucanus, Die Rätsel des Vogelzuges (1922), 121.

59. The stamp was issued around 1936.

60. J. Thienemann, Die Vogelwarte (1910), 18-19.

61. See among others: Pickstone, Ways of Knowing (2000), 151-154.

62. J. Thienemann, Rossitten (1930), 246-250.

63. Gätke, Die Vogelwarte (1891), 46-64.

64. See for instance: Von Lucanus, "Die Höhe" (1902), 1-9; idem, Die Rätsel des Vogelzuges (1922). On Lucanus's ornithological activities, Gebhart, Die Ornithologen (2006), 224.

65. J. Thienemann, Rossitten (1930), 259-262.

66. J. Thienemann, Etwas über das Experiment (1927).

67. On Schulz: Sommerfeld, Er Flug (1984).

68. J. Thienemann, Vom Vogelzuge (1931), 66.

69. J. Thienemann, Rossitten (1930), 285-300.

70. J. Thienemann, "X. Jahresbericht" (1912), 620.

71. Schüz, "Johannes Thienemann" (1938), 479.

72. Berthold and Schlenker, "Johannes Thienemann" (1995), 583.

73. Both Thienemann's father, August Wilhelm, and his grandfather, Georg August Wilhelm, were evangelical ministers known for their ornithological work. The latter often collaborated with Christian Ludwig Brehm, one of the figureheads of the "Golden Age" of German ornithology in the mid-nineteenth century and (again) a pastor. Thienemann, "Wie die alten sungen" (1936), 139-149. On the evangelical tradition of scientific research: Günther, "Pfarrer" (1984), 277-294; Daum, Wissenschaftspopularisierung (1998), 413-414.

74. The quotations come from: Livingstone, *Putting Science in Its Place* (2003), 42.

75. Tinbergen to his parents, 19 Sept. 1925, Bodleian Library, Ms. Eng. c. 3125, File 1. 14.

76. The hunting lease conflict led to a direct conflict with Thienemann's preparator, who was fired in 1916. G. Möschler to Herman Schalow, 17 Aug. 1916, Nachl. Schalow, Staatsbibliothek Berlin; Kuratorium of the *Vogelwarte* to Ministerium für Wissenschaft, Kunst und Volksbildung, 22 April 1921; Johannes Thienemann, "Kurze Denkschrift über die Verhältnisse auf der Vogelwarte Rossitten" (1921), typescript, Max-Planck-Institut für Ornithologie), Radolfzell (AMPIO).

77. J. Thienemann, *Rossitten* (1930), 3. Again, this was a common strategy among fieldworkers. Livingstone, *Putting Science in Its Place* (2003), 42.

78. J. Thienemann, "XXIII. und XXIV. Jahresbericht" (1926), 85. The phenomenon of this type of *Pfeilstorch* (or "arrow stork") was not new. In 1822, one had already been reported in the German village of Klütz. See: Kinzelbach, *Das Buch* (2005).

79. J. Thienemann, "XI. Jahresbericht" (1913), 52.

80. See, for instance, J. Thienemann, *Rossitten* (1930), 219. Pictures of the interior of the museum are kept in the archives of the Max-Planck-Institut für Ornithologie, Radolfzell (AMPIO).

81. The very gendered identities of the hunter have particularly been studied with regard to the British Empire. Yet, in other national contexts (such as Germany) rather similar identities seem to have existed. See inter alia, MacKenzie, *The Empire* (1988); Mckenzie, "'Sadly neglected'" (2005).

82. J. Thienemann, "IX. Jahresbericht"(1910), 537; idem, *Rossitten* (1930), 211-212; idem, *Das Leben* (1939), 68.

83. Chivalry toward women was also part of the codes of the club. Yet, typically, women were not allowed to be members. Thienemann, *Rossitten* (1930), 49-54.

84. This was a disappointment for Tinbergen, who started to spend more and more time with Thienemann's assistant and photographer Rudi Steinert. Tinbergen to his parents, 5 Sept. 1925, Bodleian Library, Ms. Eng. c. 3125, File 1. 14.

85. His *Rossitten* was printed four times between 1927 and 1938.

86. *Im Lande des Vogelzuges* (1928), Bundesarchiv-Filmarchiv (Berlin). On the habit of attributing individual personalities to animals through human names in nature films, see: Mitman, *Reel Nature* (1999).

87. Brochure, *Ostpreussen. Rossitten (Kurische Nehrung)* (1939), AMPIO.

88. J. Thienemann, *Die Vogelwarte* (1910), 12.

89. Worster, *Nature's Economy* (1985), 378-379.

90. J. Thienemann, "Wie die alten sungen" (1936), 172. See also: idem, *Rossitten* (1930), 32-49 and 79-82.

91. J. Thienemann to Schalow, 1 April 1912, Nachl. Schalow, Staatsbibliothek Berlin.

92. J. Thienemann, "X. Jahresbericht" (1911), 433; idem, *Rossitten* (1930), 81.

93. J. Thienemann, "Wie die alten sungen" (1936), 176.

94. Schüz, "XXV. Bericht" (1930), 105-113; Schüz, "Vogelwarte" (1932).

95. Schüz to Oskar Heinroth, 25 May 1935, AMPIO.

96. See, for instance, Schüz to Werner Rüppell, 16 Dec. 1935; Schüz to the *Notgemeinschaft*, 4 and 25 Nov. 1935; Schüz to Rudolf Drost, 31 Dec. 1936, Schüz to Heinroth, 7 Feb. 1937, AMPIO.

97. Schüz, *Vom Vogelzug* (1952), 19; Drost to Schüz and Rudolf Stadia, 3 Nov. 1936, AMPIO.

98. Max Lukas von Cranach to Heinroth, 5 Dec. 1934, AMPIO.

99. In the same documents, the lighthouse keeper with whom the station collaborated was explicitly described as "a Memelland-Lithuanian, not a Great-Lithuanian" and it was stressed that his education had been "German through and through." Heinroth to the *Notgemeinschaft*, 29 Nov. 1933; Schüz to the *Notgemeinschaft*, 1 Dec. 1934; Schüz to Heinroth, 25 May 1935; "Arbeitsplan Vogelwarte Rossitten 1938," 15 Dec. 1937, AMPIO.

100. Schüz, "Das Vogelwarte-Museum" (1934), 218.

101. Himmler had been a member of the Society of Friends of the Vogelwarte since 1935. Schüz to Karl Wolff, Jan. 1935, AMPIO; "Flugblätter für Afrika," *Der Spiegel*, 10 Jan. 1994, 53-54.

102. Tischler, *Ein Zeitbild* (1992), 46.

103. Allen, *The Naturalist in Britain* (1994), 212.

104. According to contemporaries the developments in the United States, where ringing started in the same period, were somewhat independent from those in Europe. Wood, "The History" (1945), 260.

105. In 1930 the French ornithologist (and Giardist!) Albert Chappellier listed 25 ringing stations in Europe. An important part of these were connected to museums in cities, but several others were, like the one in Rossitten, organized around a field observatory. Chappellier, "Stations de Baguage" (1930).

CHAPTER SEVEN

1. For an overview of the existing scholarship on natural history museums: Alberti, "Constructing Nature" (2008). For the topics under discussion here: Kohlstedt, "International Exchange" (1987); Allison-Bunnell, "Making Nature 'Real' Again" (1998); Thackray and Press, *The Natural History Museum* (2001); Köstering, *Natur zum Anschauen* (2003); Nyhart, "Science, Art and Authenticity" (2004); Kretschmann, *Räume öffnen sich* (2006); Nyhart, *Modern Nature* (2009), 198-250.

2. Gilson, *Le musée* (1914), 100.

3. Vivé and Versailles, *Van museum tot instituut* (1996), 9-15.

4. Forgan, "Bricks and Bones" (1999), 192.

5. On the history of the Muséum d'Histoire Naturelle: Limoges, "The Development" (1980); Blanckaert et al., *Le Muséum* (1997).

6. Diagre-Vanderpelen, *Le Jardin botanique* (2006), vol. 2, 450-455 and 585-592.

7. Vivé and Versailles, *Van museum tot instituut*, 21; Groessens and Groessens-Van Dyck, "De geologie" (2001).

8. Annual report of 1874, RBINS, Central Archive.

9. Cited in: Outram, "New Spaces" (1995), 249.

10. He outlined this idea already in the letter to the Minister of Interior in which he solicited for the job of director. Dupont to Eudore Pirmez (Minister of Interior), 3 June 1868, RBINS, Dupont papers.

11. Gilson, *Le Musée* (1914), 166-182; Vanpaemel, "Gustave Dewalque" (1996), 171-183; De Bont, *Darwins kleinkinderen* (2008), 182-192.

12. Annual reports of 1892, 1899, and 1900, RBINS, Central Archive. See also elaborately in Dupont to Frans Schollaert (Minister of Interior), 6 October 1898, State Archives, T38: VI-276: dossier Severin.

13. Dupont to Schollaert, 17 September 1898, State Archives, T38: VI-273: dossier Lebrun.

14. Among others in the annual reports of 1900, 1906, and 1911. Cf. also Guillaume Severin, "Rapport annuel sur les travaux exécutés pendant l'année 1899," manuscript, RBINS, Severin Papers.

15. On Severin: Lameere, "Guillaume Severin" (1938).

16. Annual reports of 1899, RBINS, Central Archive.

17. One of the habitués at the salon (and a good friend of Ernest Jr.) was the avant-garde painter James Ensor. Canning, "James Ensor," 29-30.

18. Lestage, "Le Dr. Ernest Rousseau" (1921).

19. According to one of his staff members, Dupont always made sure to have ideologically homogeneous dinner parties at his home, so that he could better play up to his politically influential guests: "Some days are devoted to the Catholics, other to the socialists, and still other to the liberals!" Ernest van den Broeck to Henry de Dorlodot, Dorlodot Papers, 5 March 1899, Archives de l'Université Catholique de Louvain-la Neuve.

20. Débaissieux, "Discours," 12.

21. Gilson, "Prodajus Ostendensis" (1909).

22. On the urbanistics of the quartier Leopold: Papadopoulos, *Urban Regimes* (1996), 75–77; *La morphologie spatiale* (2007).

23. Gilson, *Le musée* (1914), 85.

24. Member lists of naturalist societies indicate that, around 1900, most Brussels scientists lived in the districts of Ixelles and Etterbeek.

25. Quintart, "L'institut royal" (1990), 325.

26. See among others, Wéry, *Excursions scientifiques: Sur le littoral Belge* (1908); Massart, *Pour la protection* (1912); Stevens and Vanderswaelmen, *Guide du promeneur* (1914); Stevens, *Une excursion-type* (1915).

27. Annual reports, 1892 and 1900, RBINS, Central Archive. See also: Dupont to Jules de Trooz (Minister of Interior), 12 May 1903, State Archives, T38: VI-276: dossier Severin.

28. Gilson, *Le musée* (1914), 200.

29. Severin, "Rapport sur l'année 1912," RBINS, Severin Papers.

30. Severin, "Rapport sur l'année 1906," RBINS, Severin Papers.

31. Severin to Dupont, 25 Nov. 1895, RBINS, Severin Papers; Gilson to Paul Berryer (Minister of Interior), 31 Dec. 1923, State Archives, T38: VI-278 : Emplois temporaires.

32. Rahir, "Le nouvel aquarium" (1906); Gilson, *Le musée* (1914), 203; Rousseau, "L'Aquarium" (1917); Robeyns, "Aquarium" (2009).

33. "L'agryonète" (1920), 62.

34. Gilson, *Le musée* (1914), 25.

35. De Meyer, "Le laboratoire" (1899), 273; Photo archive Severin, RBINS. De Meyer indicates among others that Severin took up a lot of space in the small laboratory with his jars of cyanide, with which he poisoned not only insects, but also his colleagues.

36. Victor Willem to Severin, 15 Jan. 1901 and 8 Nov. 1903, Severin Papers, RBINS; Severin to Cépède 22 June 1911, Cépède Papers, Institut Pasteur.

37. Edouard Dupont to Jules de Trooz (Minister of Interior), 7 Jan. 1904, State Archives, T038-276; Gilson to Paul Berryer (Minister of Interior), 7 Jan. 1923, State Archives, T38: VI-278; Annual reports of 1911, 1912, and 1913, RBINS, Central Archive.

38. Gilson, *Le musée* (1914), 57.

39. This aquarium was founded in 1894 by the photographer Emile-Gaston Le Bon, who for that purpose had created his own society. The grotto-style aquarium was situated under Le Bon's atelier. From 1898 onward it received a yearly subsidy of the Ministry of Science—among others thanks to letters of recommendation of Gilson. Gilson to Jules de Trooz (Minister of Science), 10 Sept. 1902 and 21 March 1907, State Archives, T038-303; Gilson, *Exploration* (1900), 11; Hostyn, "Een natte attractie" (1976).

40. Gilson, *Exploration* (1900), 10; idem, *Le musée* (1914), 203–207.

41. Annual report 1907, RBINS Central Archive.

42. Rousseau, "La station biologique" (1906); idem, "Le laboratoire" (1920); Lestage, "Le Dr. Ernest Rousseau" (1921).

43. Ensch, "A propos du laboratoire ambulant" (1897), 712.

44. Severin, "Le rôle de l'entomologie" (1904).

45. Rousseau, *Notes sur la pollution* (n.d.); Rousseau, "L'Aquarium" (1917), 297.

46. Gilson, *Destruction du jeune poisson* (1932).

47. See among others: Gilson, *Exploration* (1900); idem, "L'anguille" (1908); Ernest Rousseau, *Les poissons* (1915) ; Rousseau et al., *Les larves* (1921). Annual report 1911, RBINS, Central archive.

48. Gilson, *Exploration* (1900).

49. The first fishes tagged by Gilson would be fished up in 1905: Cligny, "Poissons marqués" (1905). See also: Gilson to Jules de Trooz (Minister of Interior), 5 April 1907, State Archives, T038-289.

50. On the Commission: Rozwadowski, *The Sea* (2002).

51. Gilson, *Le musée* (1914), 75.

52. See among others: Gieryn, "City as a Truth-Spot" (2006); Kohler, "Lab History" (2008).

53. Dupont described this system as it functioned in 1869. Although he tried to reform the policy in this regard, the collection formation during his directorship continued to depend on purchase (especially when zoological collections were concerned). Annual report, 1869, RBINS, Central Archive.

54. Gilson, *Le musée* (1914), 69.

55. Annual report, 1907, RBINS, Central Archive.

56. Zimmerman, *Anthropology* (2001), 217-220.

57. Annual report, 1911, RBINS, Central Archive.

58. Gilson, *Le musée* (1914), 124-130.

59. Gilson, *Exploration* (1900), 2; Severin, "Le rôle de l'entomologie" (1904); idem, "Oiseaux" (1906); Rousseau, "Le laboratoire" (1920).

60. Rousseau, *Rapport* (1914).

61. Severin, "L'invasion" (1902).

62. Severin, "Oiseaux" (1906); Schalow, *Verhandlungen* (1911), 103.

63. This ambiguity is clear in a letter from Severin to Dupont, in which he stresses to what extent the economical issues he researched for the Commission of Waters and Forests were also of scientific importance and an integral part of the museum's ecological program. Severin to Dupont, 22 Feb. 1908, RBINS, Severin Papers.

64. Gilson, *Le musée* (1914), 116-120 and 147-148.

65. Van Straelen, "Discours" (1936), 23.

66. Edouard Dupont to Jules de Trooz (Minister of Interior), 7 Jan. 1904, State Archives, T038-276.

67. Gilson, *Le musée* (1914), 147.

68. Gilson to Berryer (Minister of Interior), 11 April 1922, State Archives, T038-289.

69. Van Straelen, "Discours" (1936), 23.

70. He would (almost literally) be described as such by his former Louvain colleague Paul Débaissieux. Débaissieux, "Discours" (1936), 16.

71. Particularly on the United States: Herzig, *Suffering for Science* (2005).

72. See among others, Wonders, *Habitat Dioramas* (1993); Nyhart, "Science, Art and Authenticity" (2004).

73. For the wider discussion of the new forms of display in fin de siecle Belgium: Nys, "Aspirations to Life" (2008).

74. Severin, "À propos d'une note" (1907); idem, "Appréciation d'un travail paru dans la Revue des questions scientifiques et concernant les musées américains," State Archives, T038-276.

75. Gilson, "Le Musée propédeutique" (1909); Nys, "Aspirations of life" (2008), 116.

76. Yet, in most places this scientific work still concerned collecting and classifying species rather than the exploration as it was conceived in Brussels. "Rapport de M. L'aide naturaliste Severin sur son voyage dans l'Europe centrale en 1898," General State Archives, Brussels, T038-276.

77. Nyhart, "Science, Art and Authenticity " (2004), 311.

78. Gilson, *Le musée* (1914), 28.

79. Dupont to Severin, 19 Nov. 1895, RBINS, Severin Papers.

80. The box in question is shown in: Gilson, *Le musée* (1914), 88. See also: Guillaume Severin, "Rapport annuel sur les travaux exécutés pendant l'année 1899," RBINS, Severin Papers.

81. Edouard Dupont to Jules de Trooz, 7 Jan. 1904, State Archives, T038-276.

82. Maybe it was those exhibition forms Severin referred to when he described the Berlin museum as too "pedagogical." "Rapport de M. L'aide naturaliste Severin," State Archives T038-276. On the exposition forms in Berlin: Kretschmann, *Räume öffnen sich* (2006), 202–210; Nyhart, *Modern Nature* (2009), 232–235.

83. See the collected photographs in the file "Mission Severin," Cartothèque, RBINS.

84. Gilson believed prehistoric man could be understood only by combining the expertise of the biologist and the stratigrapher. What his museum aimed at with its hall of prehistoric man was "an ecological study of the most elevated mammal." Gilson et al., "Musée" (1917), 84.

85. Gilson, *Le musée* (1914), 183–190; Gilson et al., "Musée" (1917), 71–141; Thiery, *Een bezoek* ([1928]); De Vos, *Museum* ([2006]).

86. Yet it used the same mechanisms as a progressively told story. Here, as well, the objects received their meaning only as part of a sequence. See: Bennett, *The Birth of the Museum* (2000), 175–208.

87. To make things easier, some visitor guides simply turned around the direction of visit. Thiery, *Een bezoek* ([1928]).

88. Gilson, *Le musée* (1914), 92–93.

89. Severin, "Rapport annuel sur les travaux exécutés pendant l'année 1904," manuscript, RBINS, Severin Papers.

90. Dupont to Jules de Trooz, 21 June 1906, State Archives, T038-291.

91. Severin, "Appréciation," State Archives, T038-276.

92. Victor Willem to Severin, 2 Nov. 1902, RBINS, Severin Papers; Annual report 1911, RBINS, Central Archive.

93. Lameere, "Laboratoire" (1896–1897).

94. See among others, Fox, "The Savant Confronts His Peers" (1980).

95. Severin, for instance, believed that a life study of insect larvae could be successful only when amateurs in the whole country were mobilized. Severin to Gilson, 25 Nov. 1895, RBINS, Severin Papers.

96. Annual report 1913, RBINS, Central Archive.

97. Dekeyser, "Notre société" (1920); Tournay, "L'histoire" (1966).

98. Dekeyser, "Notre société" (1920), 2.

99. Gilson to Pierre Nolf, 2 March 1925, State Archives, T038-290.

100. Gilson to Xavier Neujean, 9 Nov. 1921, State Archives, T038-289.

101. In his postwar reports to the Minister, Gilson lamented that his limited finances undermined an efficient functioning of the museum. In these years he even had to negotiate to get his own train season ticket paid back. Gilson to Pierre Nolf, 14 Nov. 1923, State Archives, T038-271.

102. Gilson to Léon Théodor, 16 June 1925, State Archives, T038-271.

103. In the years 1927 and 1928, Van Straelen and Gilson would engage in a bitter legal dispute about who could claim the instruments and publications that stemmed from Gilson's directorship.

104. Lameere, "L'Institut zoologique" (1927).

105. Correspondence Brussels Aquarium, State Archives, T038-303.

106. Van Straelen to François Bovesse, 18 April 1936, State Archives, T038-290.

107. Stockmans, "Notice" (1973).

108. "The Natural History Museum" (1914); Roule, "Les musées" (1923).

109. Howard, "The Recent Progress" (1912), 592.

CONCLUSION

1. Livingstone, *Putting Science in Its Place* (2003), 42.
2. Billick and Price, "The Ecology of Place" (2010).
3. Gilson, *Le musée* (1914), 75.
4. Alberti, "Amateurs and Professionals" (2001).
5. Zacharias, "Ueber die Errichtung" (1889), 23.
6. Massart, *Pour la protection* (1912), 5.
7. Zacharias, *Ueber die wissenschaftliche Bedeutung* (1905), 5.
8. The first chairs of ecology in France and Germany were respectively those of the Muséum d'Histoire Naturelle (created in 1956) and the University of Kiel (founded in 1963). Societies and journals followed in the late 1960s.

Bibliography

ARCHIVES

Berlin, Bundesarchiv-Staatsarchiv

Film: *Im Lande des Vogelzuges* (produced by Hubert Schinger, 4 reels)

Berlin, Staatsbibliothek

Nachlass Schalow: Correspondence with Johannes Thienemann

Brussels, Archives de l'Université Libre de Bruxelles

Dollo Papers: Course manuscripts
Errera Papers: Correspondence

Brussels, Royal Belgian Institute of Natural Sciences (RBINS)

Cartothèque: Photographs
Central Archives: Year Reports
Dollo Papers: Correspondence
Dupont Papers: Correspondence
Severin Papers: Correspondence, photographs

Brussels, State Archives

T38 Archives de l'administration de l'enseignement supérieur (1835–1953):
VI. Musée d'histoire naturelle

Cambridge, University Library

Darwin Papers: Correspondence with Otto Zacharias, CUL DAR 184

Kiel, Universitätsbibliothek

Nachlass Zacharias: Correspondence

Kiel, Zoologisches Museum

Nachlass Brandt: Correspondence
Nachlass Möbius: Correspondence, diaries

Leuven, Universiteitsarchief

Topografisch-historische atlas

Liège, Archives de l'Université de Liège

Dewalque Papers, 3319C: Correspondence

London, Imperial College

Huxley Correspondence, Series 1L: Letters from Lacaze-Duthiers

London, Wellcome Trust

"L'illustre Monsieur Coste," scrapbook, WMS 4 MS 1888

Louvain-la-Neuve, Archives de l'Université Catholique de Louvain

Dorlodot Papers: Correspondence

Naples, Stazione Zoologica Anton Dohrn

Correspondence Dohrn-Giard

Oxford, Bodleian Library

Correspondence Niko Tinbergen

Paris, Institut Pasteur

Fonds Casimir Cépède: Laboratory notebooks, correspondence
Fonds Maurice Caullery: Photographs

Plön, Max-Planck-Insitut für Evolutionsbiologie

August Thienemann: Guestbook, laboratory notebook, photographs
Otto Zacharias: Offprints, newspaper cuttings, photographs

Radolfzell, Max-Planck-Institut für Ornithologie (AMPIO)

Ernst Schüz: Correspondence, notebooks, photographs
Johannes Thienemann: Offprints, photographs

PUBLISHED SOURCES

Anonymous. 1838. "Notice sur les hirondelles et les autres oiseaux de passage." *Correspondance mathématique et physique de l'Observatoire* 4: 431–440.

Anonymous [Hydrographische Amt der Admiralität]. 1882. *Handbuch der nautischen Instrumente*. Berlin: Ernst Siegfried Mittler und Sohn.

Anonymous. 1883. "Marine Zoological Laboratories." *Nature* 29: 16–17.

Anonymous. 1884. "Discussion de l'interpellation de M. Giard au sujet des mines d'Anzin." *Journal officiel de la République Française: Débats parlementaires—Chambre des Députés*, 6 March, 649–654.

Anonymous. 1891. "Nouvelles scientifiques." *Revue des sciences naturelles de l'ouest* 1: 85–89.

Anonymous. 1894. *Acht fachmännische Gutachten über die Biologische Station zu Plön und deren bisherige Tätigkeit. Als Manuskript gedruckt*. Plön: S.W. Hirt's Buchdruckerei.

Anonymous. 1897. "Some Unwritten History of the Naples Zoological Station." *American Naturalist* 31: 960–965.

Anonymous. 1900–1901. "Allocution de M. Pierre Janet." *Bulletin de l'institut psychologique international* 1: 133–139.

Anonymous. n.d. [1909]. *Institut général psychologique. Notes et documents concernant l'œuvre, son objet scientifique et son caractère international, l'association et les groupements constitués, les recherches et travaux poursuivis. 1900–1908*. Paris: Institut général psychologique.

Anonymous. 1910. "The Oceanographical Museum at Monaco." *Nature* 83: 191–193.

Anonymous. 1914. "The Natural History Museum in Theory and Practice." *Nature* 94: 403–404.

Anonymous. 1920. "L'agryonète. Résumé de la causerie donnée par le Dr. Rousseau." *Les naturalistes belges* 1: 62–64.

Anonymous. 2007. *La morphologie spatiale des quartiers européens. Partie A. Diagnostic de la structure spatiale existante*. Brussels: Space Syntax.

Abel, Othenio. 1928. "Die Festgabe der 'Palaeobiologica.'" *Palaeobiologica* 1: 1–6.

———. 1928. "Louis Dollo. Zur Vollendung seines siebzigsten Lebensjahres." *Palaeobiologica* 1: 7–12.

Acloque, Alexandre, and Casimir Cépède. 1910. *Observations biologiques et écologiques sur la flore de Wimereux et de ses environs*. Boulogne sur Mer: G. Hamain.

Acot, Pascal, ed. 1998. *The European Origins of Scientific Ecology*. Paris: Editions des archives contemporaines.

Adams, Charles C. 1977 [1913]. *Guide to the Study of Animal Ecology*. New York: Macmillan.

Alberti, Samuel J. M. M. 2001. "Amateurs and Professionals in One County: Biology and Natural History in Late Victorian Yorkshire." *Journal of the History of Biology* 34: 115–145.

———. 2008. "Constructing Nature behind Glass." *Museum and Society* 6: 73–92.

Allen, David Elliston. 1994. *The Naturalist in Britain: A Social History*. Princeton: Princeton University Press.

———. 1996. "Tastes and Crazes." In *Cultures of Natural History*, edited by Nick Jardine, James A. Secord, and Emma C. Spary, 394–407. Cambridge: Cambridge University Press.

———. 2009. "Amateurs and Professionals." In *The Cambridge History of Science*. Vol. 6, *The Modern Biological and Earth Sciences*, edited by Peter J. Bowler and John V. Pickstone, 15–33. Cambridge: Cambridge University Press.

Allison-Bunnell, Steven. 1998. "Making Nature 'Real' Again: Natural History Exhibits and Public Rhetorics of Science at the Smithsonian in the Early 1960s." In *The Politics of Display: Museums, Science, Culture*, edited by Sharon MacDonald, 77–97. London: Routledge.

Anderson, Robert David. 1977. *France, 1870–1914: Politics and Society*. London: Routledge & Kegan Paul.

———. 2004. *European Universities from the Enlightenment to 1914*. Oxford: Oxford University Press.

Appel, Toby A. 1973. "Lacaze-Duthiers, Félix-Joseph Henri de." In *Dictionary of Scientific Biography*, edited by Charles Coulston Gillispie. Vol. 7, 545–546. New York: Charles Scribner's Sons.

Apstein, Carl. 1891. "Die quantitative Bestimmung des Planktons im Süsswasser." In *Die Tier- und Pflanzenwelt des Süsswassers. Einführung in das Studium derselben*, edited by Otto Zacharias. Vol. 2, 255–294. Leipzig: J. J. Weber.

———. 1893. "Antwort auf die Entgegnung des Herrn Zacharias." *Die Heimat* 3: 231–234.

———. 1893. "Veröffentlichungen aus der biologischen Station Plön." *Die Heimat* 3: 167–170.

Astarita, Tommaso. 2005. *Between Salt Water and Holy Water: A History of Southern Italy*. New York and London: W. W. Norton.

Baedeker, Karl. 1880. *Italien: Handbuch für reisenden. Dritter Theil: Unter-Italien und Sicilien*. Leipzig: Verlag von Karl Baedeker.

———. 1885. *Mittel- und Nord-Deutschland. Westlich bis zum Rhein. Handbuch für Reisende*. Leipzig: Verlag von Karl Baedeker.

Baguley, David. 2000. *Napoleon III and His Regime: An Extravaganza*. Baton Rouge: Louisiana State University Press.

Barral, Pierre. 1968. *Les fondateurs de la Troisième République*. Paris: Librairie Armand Collin.

Barrow, Mark V. 1998. *A Passion for Birds: American Ornithology after Audubon*. Princeton: Princeton University Press.

———. 2009. *Nature's Ghosts: Confronting Extinction from the Age of Jefferson to the Age of Ecology*. Chicago: University of Chicago Press.

Bashford, Dean. 1896. "Public Aquariums in Europe." *Popular Science Monthly* 50: 13–27.

Bayertz, Kurt. 1999. "Biology and Beauty: Science and Aesthetics in Fin-de-Siècle Germany." In *Fin de Siècle and its Legacy*, edited by Mikulas Teich and Roy Porter, 278–295. Cambridge: Cambridge University Press.

Beck, Naomi. 2004. "The Diffusion of Spencerism and Its Political Interpretations in France and Italy." In *Herbert Spencer: The Intellectual Legacy*, edited by Greta Jones and Robert Peel, 37–60. London: Galton Institute.

Béguet, Bruno. 1994. "La vulgarization scientifique au XIXe siècle." In *La science pour tous*, edited by Bruno Béguet, Maryline Cantor, and Ségolène Le Men, 5–48. Paris: Réunion des Musées Nationaux.

Ben-David, Joseph. 1974. "The Universities and the Growth of Science in Germany and the United States." In *Comparative Studies in Science and Society*, edited by Sal P. Restivo and Christopher K. Vanderpool, 46–81. Columbus, OH: Charles E. Merril.

Bennett, Tony. 2000. *The Birth of the Museum: History, Theory, Politics*. London: Routledge.

Benson, Keith R. 1988. "Laboratories on the New England Shore: The 'Somewhat Different Direction' of American Marine Biology." *New England Quarterly* 61: 55–78.

———. 2009. "Field Stations and Surveys." In *The Cambridge History of Science*. Vol. 6, *The Modern Biological and Earth Sciences*, edited by Peter J. Bowler and John V. Pickstone, 78–89. Cambridge: Cambridge University Press.

Berlepsch, Hans Freiherr von. 1899. *Der gesamte Vogelschutz. Seine Begründung und Ausführung* . Gera-Untermhaus: Köhler.

Bernard, Claude. 1867. *Rapport sur les progrès et la marche de la physiologie générale en France*. Paris: Imprimerie Impérial.

———. 1868. "Médecine expérimentale. II. L'expérimentation dans les sciences de la vie." *Revue des cours scientifiques* 6: 135–141.

———. 1957 [1865]. *An Introduction to the Study of Experimental Medicine*. New York: Dover.

———. 1979. *Notes pour le rapport sur les progrès de la physiologie: Manuscrit inédit présenté et commenté par M. D. Grmek*. Paris: Collège de France.

Bernardini, Jean-Marc. 1997. *Darwinisme social en France (1859–1914): Fascination et rejet d'une idéologie* . Paris: CNRS.

Berthold, Peter, and Rolf Schlenker. 1995. "Johannes Thienemann (1863–1938). Wissenschaftliche Biographie und Würdigung." In *Die Albertus-Universität zu Königsberg und ihre Professoren. Aus Anlaß der Gründung der Albertus-Universität vor 450 Jahren*, edited by Dietrich Rauschning and Donata von Nerée, 583–599. Berlin: Duncker und Humblot.

Bétencourt, Alfred. 1888. "Les hydraires du Pas-de-Calais." *Bulletin scientifique de la France et la Belgique* 19: 201–214.

Bezzenberger, Adalbert. 1889. *Die Kurische Nehrung und Ihre Bewohner*. Stuttgart: Verlag von J. Engelhorn.

Billick, Ian, and Mary V. Price. 2010. "The Ecology of Place." In *The Ecology of Place: Contributions of Place-Based Research to Ecological Understanding*, edited by Ian Billick and Mary V. Price, 1–10. Chicago: University of Chicago Press.

Blackbourn, David. 2003. *History of Germany: 1780–1980. The Long Nineteenth Century*. Oxford: Blackwell.

———. 2006. *The Conquest of Nature: Water, Landscape and the Making of Modern Germany*. London: Jonathan Cape.

Blanckaert, Claude, Claudine Cohen, Pietro Corsi, and Jean-Louis Fischer. 1997. *Le Muséum au premier siècle de son histoire*. Paris: Editions du Muséum d'histoire naturelle.

Blaringhem, Louis, Georges Bohn, Maurice Caullery, et al. 1909. "Introduction." *Bulletin scientifique de la France et de la Belgique* 43: i–ii.

Bocking, Stephen. 1997. *Ecologists and Environmental Politics: A History of Contemporary Ecology*. New Haven: Yale University Press.

Bohn, Georges. 1901. *Des mécanismes respiratoires chez les crustacés décapodes, essai de physiologie évolutive, éthologique et phylogénique*. Paris: Laboratoire d'évolution des êtres organisés.

———. 1903. "De la recherche des abris par les animaux marins littoraux." *Bulletin de l'institut général psychologique* 3: 493–495.

———. 1905. "A quoi peut-on reconnaître qu'un phénomène est 'naturel'?" *Comptes rendus des séances et mémoires de la Société de biologie* 57: 187–189.

———. 1905. *Attractions & oscillations des animaux marins sous l'influence de la lumière*. Paris: Institut général psychologique.

———. 1908. "Le passé et l'avenir de la psychologie comparée." *Revue scientifique* 9: 619–627.

———. 1909. *La naissance de l'intelligence*. Paris: Flamarion.

———. 1910. *Alfred Giard et son oeuvre*. Paris: Mercure de France.

———. 1911. *La nouvelle psychologie animale*. Paris: F. Alcan.

———. 1911. "Les progrès récents de la psychologie comparée." *L'année psychologique* 18: 478–502.

———. 1912. "La biologie générale et la psychologie comparée." *Revue scientifique* 13: 357–365.

Boissay, Charles. 1877. "Le laboratoire de zoologie maritime de Wimereux." *La nature: Revue des sciences et de leurs applications aux arts et à l'industrie* 5: 129–130.

Boulay, Nicolas. 1878. *Révision de la Flore des départements du nord de la France*. Lille: L. Quarré and F. Savy.

Boutan, Louis. 1893. "Mémoire sur la photographie sous-marine." *Archives de zoologie expérimentale et générale* ser. 3, 1: 281-324.

———. 1900. La photographie sous-marine et les progrès de la photographie. Paris: Schleicher et Frères.

Bouyssi, François. 1999. *Alfred Giard (1846-1908) et ses élèves: Un cénacle de "philosophes biologistes."* Aux origines du scientisme. PhD diss., École pratique des hautes études, Paris.

Boveri, Theodor. 1910. *Anton Dohrn: Gedächtnisrede gehalten auf dem Internationalen Zoologen-Kongress in Graz am 18. August 1910.* Leipzig: Verl. von S. Hirzel.

Bowler, Peter. 1992. *The Earth Encompassed: A History of the Environmental Sciences.* New York: Norton.

———. 1996. *Life's Splendid Drama: Evolutionary Biology and the Reconstruction of Life's Ancestry, 1860-1940.* Chicago: University of Chicago Press.

Bowler, Peter, and Iwan Rhys Morus. 2005. *Making Modern Science: A Historical Survey.* Chicago: University of Chicago Press.

Brandt, Otto. 1981. *Geschichte Schleswig-Holsteins: Ein Grundriss.* Kiel: Walter G. Mühlau Verlag.

Braun, Marta. 1992. *Picturing Time: The Work of Etienne-Jules Marey (1830-1904).* Chicago: University of Chicago Press.

Brien, Paul. 1951. "Notice sur Louis Dollo." *Annuaire de l'Académie royale de Belgique* 117: 69-138.

Brower, Brady. 2010. *Unruly Spirits: The Science of Psychic Phenomena in Modern France.* Urbana: University of Illinois Press.

Bruce, Robert V. 1987. *The Launching of Modern American Science 1846-1876.* New York: Alfred A. Knopf.

Brunner, Bernd. 2003. *The Ocean at Home: An Illustrated History of the Aquarium.* New York: Princeton Architectural Press.

Bruyant, C. 1900. "La station limnologique de Besse." *Revue internationale de l'enseignement* 119.

Buchholz, Wilhelmine [Julius Stinde]. 1884. *Buchholzens in Italien. Reisen-Abenteuer von Wilhelmine Buchholz.* Berlin: Verlag von Freund & Jedel.

Burkhardt, Richard W., Jr. 2005. *Patterns of Behavior: Konrad Lorenz, Niko Tinbergen and the Foundation of Ethology.* Chicago: University of Chicago Press.

Busch, Alexander. 1962. "The Vicissitudes of the *Privatdozent*: Breakdown and Adaptation in the Recruitment of the German University Teacher." *Minerva: A Review of Science Learning and Policy* 1: 319-341.

Büschenfeld, Jürgen. 1997. *Flüsse und Kloaken: Umweltfragen im Zeitalter der Industrialisierung (1870-1918).* Stuttgart: Clett-Cotta.

Canning, Susan. 2009. "James Ensor: Carnival of the Modern." In *James Ensor*, edited by Anna Swinbourne, 28-44. New York: Museum of Modern Art.

Capus, Guillaume. 1903. *Guide du naturaliste préparateur et du voyageur scientifique. Troisième édition entièrement refondue par le Dr. Georges Bohn. Introduction par Edmond Perrier.* Paris: J. B. Baillière.

Cartaz, Adolphe. 1900. "La station zoologique de Wimereux." *La nature: Revue des sciences et de leurs applications aux arts et à l'industrie* 28: 146-148.

Caullery, Maurice. 1905. "Les yeux et l'adaptation au milieu chez les animaux abyssaux." *Revue générale des sciences pures et appliquées* 16: 324-340.

———. 1909. "L'oeuvre scientifique d'Alfred Giard." *Bulletin scientifique de la France et de la Belgique* 42: xv-xxviii.

———. 1950. "Les stations françaises de biologie marine." *Notes and Records of the Royal Society* 8: 95-115.

Cépède, Casimir. 1908. "Les manuscrits de Lamarck." *Revue des idées* 5: 257-263.

———. 1910. "Recherches sur les infusoires astomes." *Archives de zoologie expérimentale et générale* 3: 341-609.

———. 1912. *Convois migrateurs estivaux des libellules (L. Quadrimaculata L.)*.Boulogne-sur-Mer: G. Hamain.

Chappellier, Albert. 1930. "Stations de Baguage." *Revue d'Histoire Naturelle (2e partie): l'Oiseau et la Revue française d'ornithologie* 11: 342–350.

Chevreul, Michel E. 1870. *De la méthode à posteriori expérimentale et de la généralité de ses applications*. Paris: Dunod.

Chun, Carl. 1880. *Die Ctenophoren des Golfes von Neapel und der angrenzende Meeres-Abschnitte: Eine Monographie*. Leipzig: Wilhelm Engelmann.

Chun, Carl, and Wilhelm Johannsen, eds. 1915. *Allgemeine Biologie*. Berlin: B. G. Teubner.

Cittadino, Eugene. 1990. *Nature as the Laboratory: Darwinian Plant Ecology in the German Empire, 1880–1900*. Cambridge: Cambridge University Press, 1990.

Clark, Terry Nichols. 1973. *Prophets and Patrons: The French University and the Emergence of the Social Sciences*. Cambridge, MA: Harvard University Press.

Cligny, Adolphe. 1904. "Sur l'éthologie du hareng." *Comptes rendus des séances de la Société de Biologie* 57: 347–349.

———. 1905. "Poissons marqués." *Annales de la station aquicole de Boulogne-sur-Mer* 1: 123–126.

Coleman, William. 1985. "The Cognitive Basis of the Discipline: Claude Bernard on Physiology." *Isis* 76: 49–70.

———. 1986. "Evolution into Ecology? The Strategy of Warming's Ecological Plant Geography." *Journal of the History of Biology* 19: 181–196.

Coleman, William, and Frederic L. Holmes, eds. 1988. *The Investigative Enterprise: Experimental Physiology in Nineteenth-Century Medicine*. Berkeley: University of California Press.

Conry, Yvette. 1974. *L'introduction du darwinisme en France au XIXe siècle*. Paris: Vrin.

Coste, Victor. 1848. *Nidification des épinoches et des épinochettes*. Paris: Imprimerie nationale.

———. 1853. "Mémoire sur les moyens de repeupler les eaux de la France." *Comtes rendus de l'Académie des sciences* 36: 237–245.

———. 1869. *De l'observation et de l'expérience en physiologie*. Paris: Masson.

Coulon, Marcel. 1925. *Les ennemis de J.-H. Fabre et Ferton*. Paris: Les éditions du monde moderne.

Coupin, Henri. 1894. "Le laboratoire maritime de Saint-Vaast-de-la-Hougue." *La nature: Revue des sciences et de leurs applications aux arts et à l'industrie* 22: 343–346.

Crombois, Jean-François. 1994. *L'univers de la sociologie en Belgique de 1900 à 1940*. Brussels: ULB.

Crowfort, Peter. 1992. *Elton's Ecologists: A History of the Bureau of Animal Population*. Chicago: University of Chicago Press.

Cunningham, Joseph. 1884. "La station maritime de Granton-Édimbourg." *Archives de zoologie expérimentale et générale* ser. 2, 2: xvii–xix.

Cuvier, Georges. 1837. *The Animal Kingdom arranged according to its Organization serving as a Foundation for the Natural History of Animals and an Introduction to Comparative Anatomy*. London: G. Henderson. 4 vols.

Dahl, Friedrich. 1898. "Experimentell-Statistische Ethologie." *Verhandlungen der Deutschen Zoologischen Gesellschaft* 8: 121–131.

———. 1901. "Was ist ein Experiment, was Statistik in der Ethologie?" *Biologische Centralblatt* 21: 675–681.

———. 1902. "Die Zieler der vergleichende 'Ethologie' (d.i. Biologie im älteren engeren Sinne)" In *Verhandlungen des V. Internationalen Zoologen-Congresses zu Berlin, 12.-16. August 1901*. 296–300. Jena: Fischer.

Daled, Pierre F. 1998. *Spiritualisme et matérialisme au XIXe siècle: L'université libre de Bruxelles et la religion*. Brussels: Editions de l'Université de Bruxelles.

Darby, Margaret Flanders. 2007. "Unnatural History: Ward's Glass Cases." *Victorian Literature & Culture* 35: 635–647.

Daston, Lorraine, and Peter Galison. 2010. *Objectivity*. New York: Zone Books.

Daston, Lorraine, and Elisabeth Lunbeck. 2011. "Introduction." In *Histories of Scientific Observation*, edited by Lorraine Daston and Elizabeth Lunbeck, 1–9. Chicago: University of Chicago Press.

Daum, Andreas. 1995. "Naturwissenschaftlicher Journalismus im Dienst der darwinistischen Weltanschauung: Ernst Krause, alias Carus Sterne, Ernst Haeckel und die Zeitschrift Kosmos. eine Fallstudie zum späten 19. Jahrhundert." *Mauritiania* 15: 227–245.

———. 1998. *Wissenschaftspopularisierung im 19. Jahrhundert: Bürgerliche Kultur, naturwissenschaftliche Bildung und die deutsche Öffentlichkeit, 1848–1914.* Munich: R. Oldenbourg.

———. 2002. "Science, Politics and Religion: Humboldtian Thinking and the Transformations of Civil Society in Germany, 1830–1870." In *Science and Civil Society*, edited by Lynn K. Nyhart and Thomas Broman. *Osiris* 17: 107–140.

Deacon, Margaret. 1997. *Scientists and the Sea, 1650–1900: A Study of Marine Science.* London: Ashgate.

Débaissieux, Paul. 1936. "Discours de M. P. Débaissieux." In *Manifestation en l'honneur de monsieur Gustave Gilson, professeur à l'Université catholique de Louvain à l'occasion de son jubilé professoral: 1885–1936*, 15–21. Louvain: Université Catholique de Louvain.

Debaz, Josquin. 2005. *Les stations françaises de biologie marine et leurs périodiques entre 1872 et 1914.* PhD diss., École des hautes études en sciences sociales, Paris.

De Bont, Raf. 2008. "Evolutionary Morphology in Belgium: The Fortunes of the Van Beneden School." *Journal of the History of Biology* 41: 81–118.

———. 2008. *Darwins kleinkinderen. De evolutietheorie in België, 1865–1945.* Nijmegen: Vantilt.

De Bont, Raf, and Rajesh Heynickx. 2012. "Landscapes of Nostalgia: Biologists and Literary Intellectuals Protecting Belgium's 'Wilderness.'" *Environment and History* 18: 237–260.

De Fonvielle, Wilfrid. 1873. "Coste." *La nature: Revue des sciences et de leurs applications aux arts et à l'industrie* 1: 295–297 and 327–329.

De Guerne, Jules. 1892. "Le laboratoire de biologie du Lac de Ploen." *Revue biologique du Nord de la France* 4: 146–149.

Dekeyser, Léon. 1920. "Notre société." *Les naturalistes belges* 1: 1–4.

De Lacaze-Duthiers, Henri. 1872. "Avertissement." *Archives de zoologie expérimentale et générale* 1: v–vii.

———. 1872. "Création d'un laboratoire de zoologie expérimentale sur les côtes de la France." *Archives de zoologie expérimentale et générale* 1: I–lii.

———. 1872. "Direction des études zoologiques." *Archives de zoologie expérimentale et générale* 1: 1–64.

———. 1886. "Dix-sept années d'enseignement de la zoologie en Sorbonne." *Revue scientifique* 23: 737–748.

———. 1898. "Sur les laboratoires de Roscoff, Banyuls et les Archives." *Archives de zoologie expérimentale et générale*, ser. 3, 6: 1–35.

———. 1967 [1872]. "The Study of Zoology—On Experimental Procedure in the Natural Sciences and Particularly in Zoology." In *The Interpretation of Animal Form*, edited by William Coleman, 135–163. New York: Johnson.

Demarée, Gaston René, and Isabelle Chuine. 2006. "A Concise History of the Phenological Observations at the Royal Meteorological Institute of Belgium." In *HAICTA 2006*. 3: 815–824.

De Meyer, Jean. 1899. "Le laboratoire ambulant de biologie. Excursion de Francorchamps." *Revue de l'Université de Bruxelles* 5: 272–277.

Demoor, Jean, Jean Massart, and Émile Vandervelde. 1897. *L'évolution régressive en biologie et en sociologie.* Paris: Alcan.

De Parville, Henri. 1901. "Henri de Lacaze-Duthiers." *La nature: Revue des sciences et de leurs applications aux arts et à l'industrie* 29: 159–160.

De Quatrefages de Bréau, Armand. 1862. *Fertilité et culture d'eau.* Paris: Société impériale zoologique d'acclimatation.

Desmond, Adrian. 1979. "Designing the Dinosaur: Richard Owen's Response to Robert Edward Grant." *Isis* 70: 224–234.

De Vos, Wim. [2006]. *Museum voor natuurwetenschappen. De Janletzaal/Een vleugel van licht voor de dinosauriërs*. Antwerp: Openbaar Kunstbezit Vlaanderen.

Diagre-Vanderpelen, Denis. 2006. *Le Jardin botanique de Bruxelles (1826–1912). Miroir d'une jeune nation*. PhD diss., Free University of Brussels, Brussels. 2 vols.

Dierig, Sven. 2003. "Engines for Experiment: Laboratory Revolution and Industrial Labor in the Nineteenth-Century City." In *Science in the City*, edited by Sven Dierig, Jens Lachmund, and Andrew Mendelsohn. *Osiris* 18: 116–134.

———. 2006. *Wissenschaft in der Machinenstadt: Emil Du Bois-Reymond und seine Laboratorien in Berlin*. Göttingen: Wallstein Verlag.

Dierig, Sven, Jens Lachmund, and Andrew Mendelsohn, eds. 2003. *Science in the City*. Chicago: University of Chicago Press. *Osiris* 18.

Di Gregorio, Mario. 2005. *From Here to Eternity: Ernst Haeckel and Scientific Faith*. Göttingen: Vandenhoeck & Ruprecht.

Dohrn, Anton. 1871. *Kurzer Abriss der Geschichte sowie Gutachten und Meinungsäusserungen hervorragender Naturforscher über die Griindung zoologischer Stationen*. Jena: Fromann.

———. 1872. "The Foundation of Zoological Stations." *Nature* 5: 277–280 and 437–440.

———. 1875. *Der Ursprung der Wirbelthiere und das Princip des Functionswechsels. Genealogische Skizzen*. Leipzig: Wilhelm Engelmann.

———. 1876. *Erster Jahresbericht der Zoologischen Station in Neapel*. Leipzig: Wilhelm Engelmann.

———. 1879. "Bericht über die zoologische Station während der Jahre 1876–1878." *Mittheilungen aus der Zoologischen Station zu Neapel* 1: 137–164.

———. 1881. "Bericht über die zoologische Station während der Jahre 1879–1880." *Mittheilungen aus der Zoologischen Station zu Neapel* 2: 495–514.

———. 1882. "Bericht über die zoologische Station während des Jahres 1881." *Mittheilungen aus der Zoologischen Station zu Neapel* 3: 591–602.

———. 1886. "Bericht über die zoologische Station während der Jahre 1882–1884." *Mittheilungen aus der Zoologischen Station zu Neapel* 6: 93–148.

———. 1893. "Bericht über die zoologische Station während der Jahre 1885–1892." *Mittheilungen aus der Zoologischen Station zu Neapel* 10: 633–674.

———. 1926 [1872]. "Der gegenwärtige Stand der Zoologie und die Gründung zoologischer Stationen." *Die Naturwissenschaften* 44: 412–424.

Dollo, Louis. 1883. "Troisième note sur les dinosauriens de Bernissart." *Bulletin du Musée royal d'histoire naturelle de Belgique* 2: 85–119.

———. 1893. "Les lois de l'évolution." *Bulletin de la Société belge de géologie, de paléontologie et d'hydrologie* 7: 164–166.

———. 1909. "La paléontologie éthologique." *Bulletin de la Société belge de géologie, de paléontologie et d'hydrologie* 23: 377–421.

Doughty, Robin W. 1975. *Feather Fashions and Bird Preservation: A Study in Nature Protection*. Berkeley: University of California Press.

Du Bois-Reymond, Emil. 1885. "Lebende Zitterrochen in Berlin: Zweite Mittheilung." 691–750.

Du Bois-Reymond, Estelle, ed. 1982. *Two Great Scientists of the Nineteenth Century: Correspondence of Emil Du Bois-Reymond and Carl Ludwig*. Baltimore: Johns Hopkins University Press.

Durant, John R. 1981. "Innate Character in Animals and Man: A Perspective on the Origins of Ethology." In *Biology, Medicine and Society 1840–1940*, edited by Charles Webster, 157–192. Cambridge: Cambridge University Press.

Eckart, Wolfgang U. 1992. *Wissenschaft und Stadt*, a special issue of *Berichte zur Wissenschaftsgeschichte* 15: 69–225.

Edwards, Charles Lincoln. 1910. "The Zoological Station at Naples." *Popular Science Monthly* 77: 211–225.

Egerton, Frank N. 1962. "The Scientific Contributions of François Alphonse Forel, the Founder of Linology." *Schweizerische Zeitschrift für Hydrologie* 24: 181-199.

———. 1976. "Ecological Studies and Observations before 1900." In *Issues and Ideas in America*, edited by B. J. Taylor and T. J. White, 311-351. Norman: University of Oklahoma Press.

———. 2008. "Thienemann, August Friedrich." In *New Dictionary of Scientific Biography*, edited by Noretta Koertge. Vol. 7. 35-39. Detroit: Scribner's.

Ehrenbaum, Ernst. 1889. "Zoologische Wanderstation an der Nordsee." *Mittheilungen der Section für Küsten und Hochseefischerei* 4-10.

Eigen, Edward. 2001. "Dark Space and the Early Days of Photography as a Medium." *Grey Room* 3: 90-110.

———. 2003. "The Place of Distribution: Episodes in the Architecture of Experiment." In *Architecture and the Sciences: Exchanging Metaphors*, edited by Antoine Picon and Allessandra Ponte. 52-79. Princeton: Princeton Architectural Press.

Elton, Charles. 1927. *Animal Ecology*. London: Macmillan.

Ensch, Norbert. 1897. "A propos du laboratoire ambulant." *Revue de l'Université de Bruxelles* 1: 712.

Ensch, Norbert, and Louis Querton. 1897. "La station zoologique de Wimereux." *Revue de l'Université de Bruxelles* 1: 305-310.

Erlingsson, Steindor J. 2009. "The Plymouth Laboratory and the Institutionalization of Experimental Zoology in Britain in the 1920s." *Journal of the History of Biology* 42: 151-183.

Fantini, Bernardino. 2000. "The 'Stazione Zoologica Anton Dohrn' and the History of Embryology." *International Journal of Developmental Biology* 44: 523-535.

Farber, Paul L. 1999. "French Evolutionary Ethics during the Third Republic: Jean de Lanessan." In *Biology and the Foundation of Ethics*, edited by Jane Maienschein and Michael Ruse. 84-97. Cambridge: Cambridge University Press.

Fauré-Fremiet, Emmanuel. 1960. *Notice sur la vie et les travaux de Victor Coste, 1807-1873*. Paris: Palais de l'Institut.

Finnegan, Diarmid A. 2008. "The Spatial Turn: Geographical Approaches to the History of Science." *Journal of the History of Biology* 41: 369-388.

Floericke, Kurt. 1894. "Die Gründung einer ornithologische Station in Rossitten." *Die gefiederte Welt: Wochenschrift für Vogelliebhaber, -Züchter und Händler* 13: 233-235.

———. 1902. *Kritik der Tätigkeit der Vogelwarte Rossitten*. Wien: Carl Fischer.

———. 1908. *Jahrbuch der Vogelkunde*. Vol. 1. Die Forschungsergebnisse und Fortschritte der paläarktischen Ornithologie im Jahre 1907. Stuttgart: Kosmos.

Florey, Ernst. 1995. "Highlights and Sidelights of Early Biology in Helgoland." In *The Challenge to Marine Biology in a Changing World. Helgoland, 13th-18th September 1992*, edited by H.D. Franke and K. Lüning. 77-101. A special issue of *Helgoländer Meeresunteruschungen*, 49.

Florkin, Marcel, and Jean Théodoridès, eds. 1982. "Henri de Lacaze-Duthiers et Léon Frédericq: Correspondance (1876-1900)." *Archives internationales de physiologie et de biochimie* 90: 1-94.

Forbes, Stephen A. 1887. "The Lake as a Microcosm." *Bulletin of the Peoria Scientific Association* 77-87.

Forgan, Sophie. 1999. "Bricks and Bones: Architecture and Science in Victorian Britain." In *The Architecture of Science* edited by Peter Galison en Emily Thompson. 181-208. Cambridge, Mass. and London: MIT Press.

Fox, Robert. 1976. "Scientific Enterprise and the Patronage of Research in France 1800-70." In *The Patronage of Science in the Nineteenth Century*, edited by G. L. E. Turner. 9-51. Leyden: Noordhoff International Publishing.

———. 1980. "The Savant Confronts His Peers: Scientific Societies in France, 1815-1914." In *The Organization of Science and Technology in France, 1804-1914*, edited by Robert Fox and George Weisz. 241-182. Cambridge: Cambridge University Press.

———. 1984. "Science, the University and the State in Nineteenth-Century France." In *Professions and the French State, 1700-1900*, edited by Gerald L. Geison. 66-145. Philadelphia: University of Pennsylvania Press.

————. 2012. *The Savant and the State: Science and Cultural Politics in Nineteenth-Century France.* Baltimore: The John Hopkins University Press.

Francé, Raoul. 1906. *Das Leben der Pflanze.* 4 vols.

————. 1907. *Der Bildungswert der Kleinwelt: Gedanken über mikroskopische Studien.* Stuttgart: Franckh'sche Verlagshandlung.

Franke, Ulrich. 2009. *Dr. Curt Floericke: Naturforscher, Ornithologe, Schriftsteller.* Norderstedt: Books on Demand.

Frič, Antonín, and Václav Vavrá. 1894. *Die Thierwelt des Unterpočernitzer und Gatterschlagers Teiches als Resultat der Arbeiten an der Übertragbaren Zoologischen Station.* Prag: Archiv der Naturwissenschaftl. Landesdurchforschung von Böhmen.

Frost, Henry H. 1960. *The Functional Sociology of Émile Waxweiler and the Institut de Sociologie Solvay.* Brussels: Palais des Académies.

Galison, Peter. 1997. *Image and Logic: A Material Culture of Microphysics.* Chicago: University of Chicago Press.

————. 1999. "Buildings and the Subject of Science." In *The Architecture of Science,* edited by Peter Galison and Emily Thompson. 1-21. Cambridge, Mass. and London: MIT Press.

Gätke, Heinrich. 1891. *Die Vogelwarte Helgoland.* Braunschweig: Joh. Heinrich Meyer.

Gebhart, Ludwig. 2006. *Die Ornithologen Mitteleuropas. 1747 bemerkenswerte Biographien vom Mittelalter bis zum Ende des 20. Jahrhunderts* Giesen: AULA Verlag.

Geoffroy Saint-Hilaire, Isidore. 1854-1862. *Histoire naturelle des règnes organiques.* 3 vols. Paris: Victor Masson et Fils.

Ghiselin, Michael T. 2003. "Carl Gegenbaur versus Anton Dohrn." *Theory of the Biosciences* 122: 142-147.

Ghiselin, Michael T., and Christiane Groeben. 1997. "Elias Metschnikoff, Anton Dohrn, and the Metazoan Common Ancestor." *Journal of the History of Biology* 30: 211-228.

————. 2000. "A Bioeconomic Perspective on the Organization of the Naples Zoological Station." In *Cultures and Institutions of Natural History: Essays in the History and Philosophy of Science,* edited by Michael T. Ghiselin and Alan E. Leviton. 273-285. San Francisco: California Academy of Sciences.

Giard, Alfred. 1891. "Sur le bourgeonnement des larves d'Astellium spongiforme Gd. et sur la poecilogonie chez les Ascidies composées." *Comptes rendus des séances de l'Académie des sciences de l'Institut de France* 112: 301-304.

————. 1894. "Convergence and poecilonogy among Insects." *Psyche* 7: 171-175.

————. 1896 [1876]. "Considérations sur les insectes parasites à l'occassion de la chrysomèle de la pomme de terre Doryphora (Leptinotarsa) decemlineata." In *Exposé des titres et travaux scientifiques (1869-1896).* 343-344. Paris: Imprimerie générale Lahure.

————. 1896 [1881]. "Deux ennemis de l'ostréiculture." In *Exposé des titres et travaux scientifiques (1869-1896).* 362. Paris: Imprimerie générale Lahure.

————. 1896 [1888]. "Nouvelles remarques sur le Silpha opaca L." In *Exposé des titres et travaux scientifiques (1869-1896),* 347-348. Paris: Imprimerie générale Lahure.

————. 1896 [1888]. "Les saumons de la Canche." In *Exposé des titres et travaux scientifiques (1869-1896).* 358-359. Paris: Imprimerie générale Lahure.

————. 1896 [1888]. "Sur le Peroderma cylindricum Heller: Copépode parasite de la Sardine." In *Exposé des titres et travaux scientifiques (1869-1896).* 212. Paris: Imprimerie générale Lahure.

————. 1896 [1888]. "Sur le Silpha opaca L., insecte destructeur de la betterave." In *Exposé des titres et travaux scientifiques (1869-1896).* 345-347. Paris: Imprimerie générale Lahure.

————. 1896 [1889]. "Sur un convoi migrateur de Libellula quadrumaculata L. dans le nord de la France." In *Exposé des titres et travaux scientifiques (1869-1896).* 219-220. Paris: Imprimerie générale Lahure.

————. 1904 [1889]. "Les facteurs de l'évolution." In *Controverses transformistes.* 109-134. Paris: Masson.

———. 1904 [1890]. "Le principe de Lamarck et l'hérédité des modifications somatiques." In *Controverses transformistes*. 135-158. Paris: Masson.

———. 1905. "Les tendances actuelles de la morphologie et ses rapports avec les autres sciences." *Revue scientifique (revue rose)* ser. 5, 4: 29-136 and 166-172.

———. 1905. "L'évolution des sciences biologiques." *Revue scientifique (revue rose)* ser. 5, 4: 193-205.

———. 1911 [1878]. "Particularités de reproduction de certains échinodermes en rapport avec l'éthologie de ces animaux." In *Oeuvres diverses*. Vol. 1. 512-513. Paris: Laboratoire de l'Évolution des Êtres Organisés.

———. 1911 [1887], "La castration parasitaire et son influence sur les caractères extérieurs du sexe mâle chez les Crustacés Décapodes. " In *Oeuvres diverses*. Vol. 1. 241-262. Paris: Laboratoire de l'Évolution des Êtres Organisés.

———. 1911 [1889]. "De l'influence de l'éthologie de l'adulte sur l'ontogénie du Palaemonetes Varians Leach." In *Oeuvres diverses*. Vol. 1., 397-399. Paris: Laboratoire de l'Évolution des Êtres Organisés.

———. 1911 [1889]. "Sur la transformation de Publicaria Dysenterica Gaertn. en une plante di-oïque [1889]." In *Oeuvres diverses*. Vol. 1, 311-331. Paris: Laboratoire de l'Évolution des Êtres Organisés.

———. 1911 [1896]. "La méthode expérimentale en entomologie." In *Oeuvres diverses*. Vol. 1, 195-201. Paris: Laboratoire de l'Évolution des Êtres Organisés.

———. 1911 [1896]. "L'anhydrobiose ou ralentissement des phénomènes vitaux sous l'influence de la déshydratation progressives." In *Oeuvres diverses*. Vol. 1, 241-262. Paris: Laboratoire de l'Évolution des Êtres Organisés.

———. 1911 [1896]. "Préface de la notice sur les titres et travaux scientifiques." In *Oeuvres diverses*. Vol. 1, 3-40. Paris: Laboratoire de l'Évolution des Êtres Organisés.

———. 1911 [1896]. "Sur le mimétisme et la ressemblance protectrice." In *Oeuvres diverses*. Vol. 1, 479-483. Paris: Laboratoire de l'Évolution des Êtres Organisés.

———. 1911 [1905]. "Les origines de l'amour maternel." In *Oeuvres diverses*. *Vol.* 1, 207-235. Paris: Laboratoire de l'Évolution des Êtres Organisés.

———. 1911 [1908]. "L'éducation du morphologiste." In *Oeuvres diverses*. *Vol.* 1, 41-56. Paris: Laboratoire de l'Évolution des Êtres Organisés.

———. 1911 [1908]. "Un amphipode mimétique des hydraires: Metopa Rubrovittata G.O. Sars." In *Oeuvres diverses*. Vol. 1, 489-490. Paris: Laboratoire de l'Évolution des Êtres Organisés.

———. 1913 [1874]. "Discours prononcé à l'inauguration du laboratoire de zoologie maritime à Wimereux (Pas-de-Calais)." In *Oeuvres diverses*. Vol. 2, 5-17. Paris: Laboratoire de l'Évolution des Êtres Organisés.

———. 1913 [1888]. "Le laboratoire de Wimereux en 1888 (Recherches fauniques)." In *Oeuvres diverses*. Vol. 2, 18-37. Paris: Laboratoire de l'Évolution des Êtres Organisés.

———. 1913 [1889]. "Le laboratoire de Wimereux en 1889 (Recherches fauniques)." In *Oeuvres diverses*. Vol. 2, 38-62. Paris: Laboratoire de l'Évolution des Êtres Organisés.

———. 1913 [1889]. "Le laboratoire du Portel. Les grandes et les petites stations maritimes." In *Oeuvres diverses*. Vol. 2, 63-74. Paris: Laboratoire de l'Évolution des Êtres Organisés.

———. 1913 [1899]. "Coup d'oeuil sur la faune du Boulonnais." In *Oeuvres diverses*. Vol. 2, 90-146. Paris: Laboratoire de l'Évolution des Êtres Organisés.

———. 1913 [1904]. "Sur l'habitat de Silene Maritima Wither. Dans le Nord de la France." In *Oeuvres diverses*. Vol. 2, 252-253. Paris: Laboratoire de l'Évolution des Êtres Organisés.

Giard, Alfred, and Jules Bonnier. 1887. *Contributions à l'Étude des Bopyriens*. Lille: L. Danel.

Giard, Alfred, and Georges Roché. 1896 [1895]. "Rapport adressé au Ministère de la Marine au nom du Comité consultatif des pêches maritimes." In *Exposé des titres et travaux scientifiques (1869-1896)*, 360-362. Paris: Imprimerie générale Lahure.

Giard, Alfred, and Alfred-Victor Roussin. 1896 [1889]. "Comité consultatif des pêches maritimes—Rapports adressés au Ministre de la Marine." In *Exposé des titres et travaux scientifiques (1869-1896)*, 359-360. Paris: Imprimerie générale Lahure.

Gieryn, Thomas. 1998. "Biotechnology's Private Parts (and Some Public Ones)." In *Making Space for Science: Territorial Themes in the Shaping of Knowledge*, edited by Crosbie Smith and Jon Agar, 281-312. Basingstoke: Macmillan.

———. 2002. "Three Truth-Spots." *Journal of the History of the Behavioral Sciences* 38: 113-132.

———. 2006. "City as a Truth-Spot." *Social Studies of Science* 36: 5-38.

Gilson, Gustave. 1900. *Exploration de la mer sur les côtes de la Belgique en 1899*. Brussels: Polleunis & Ceuterick.

———. 1908. "L'anguille, sa reproduction, ses migrations et son intérêt économique en Belgique." *Annales de la Société royale de zoologie et de malacologie de Belgique* 42: 7-58.

———. 1909. "Le Musée propédeutique. Essai sur la création d'un organisme éducatif extra-scolaire." *Annales de la Société royale zoologique et malacologique de Belgique* 43: 46-62.

———. 1909. "*Prodajus Ostendensis* n. sp. Etude monographique d'un épicarde parasite du *Gastrosaccus Spinifer* Goes." *Bulletin scientifique de la France et de la Belgique* 43: 19-92.

———. 1914. *Le musée d'histoire naturelle moderne; Sa mission, son organization, ses droits*. Brussels: Musée d'histoire naturelle.

———. 1932. *Destruction du jeune poisson par la pêche littorale à moteur.* Ostende: Institut d'Etudes Maritimes.

Gilson, Gustave, Louis Dollo, and Aimé Rutot. 1917. "Musée Royale d'Histoire Naturelle." In *Guide Illustré de Bruxelles: Les musées*, edited by Guillaume Des Marez, 71-145. Brussels: Touring Club de Belgique.

Girod, Paul. 1893. "La station biologique des Monts Dore d'Auvergne." In *Association française pour l'avancement des sciences fusionnée avec l'Association scientifique de France. Compte rendu de la 22e session*. Paris: Association française pour l'avancement des sciences, 255.

Gissis, Snait. 2002. "Late Nineteenth Century Lamarckism and French Sociology." *Perspectives on Science* 10: 69-122

Goldman, Harvey. 1992. *Politics, Death and the Devil: Self and Power in Max Weber and Thomas Mann*. Berkeley: University of California Press.

Golinski, Jan. 1998. *Making Natural Knowledge: Constructivism and the History of Science*. Cambridge: Cambridge University Press.

Gosse, Philip Henry. 1853. "Sea-side Recreations." *New Monthly Magazine* 9: 298-305.

Gould, Stephen J. 1970. "Dollo on Dollo's Law: Irreversibility and the Status of Evolutionary Laws." *Journal of the History of Biology* 3: 189-212.

Groeben, Christiane, ed. 1982. *Charles Darwin 1809-1882 Anton Dohrn 1840-1909: Correspondence*. Naples: Macchiaroli.

———. 1985. "Anton Dohrn—The Statesman of Darwinism." *Biological Bulletin* 168: 4-25.

———. 1990. "The Vettor Pisani Circumnavigation (1882-1885)." In *Ocean Sciences: Their History and Relation to Man*, edited by Walter Lenz and Margaret Deacon, 220-233. Hamburg: Bundesamt für Seeschiffahrt und Hydrographie.

———, ed. 1993. *Correspondence: Karl Ernst von Baer (1792-1876)—Anton Dohrn (1840-1909)*. Philadelphia: American Philosophical Society.

———. 2002. "The Stazione Zoologica: A Clearing House for Marine Organisms." In *Oceanographic History: The Pacific and Beyond. Proceedings of the Fifth International Congress on the History of Oceanography, La Jolla, California, July 1993*, edited by Keith Benson and Philip F. Rehbock, 537-547. Seattle: University of Washington Press.

———. 2006. "The Stazione Zoologica Anton Dohrn as a Place for the Circulation of Scientific Ideas: Vision and Management." In *Information for Responsible Fisheries: Libraries as Mediators*, edited by Kristin L. Anderson and Cécile Thiéry, 291-299. Fort Pierce, FL: IAMSLIC.

———. 2008. "Tourists in Science: 19th Century Research Trips to the Mediterranean." *Proceedings of the California Academy of Sciences* 59: 139-154.

Groeben, Christiane, and Klaus Wenig, eds. 1992. *Anton Dohrn and Rudolf Virchow: Briefwechsel 1864-1902.* Berlin: Akademie Verlag.

Groeben, Christiane, Klaus Hierholzer, and Ernst Florey, eds.. 1985. *Emil du Bois-Reymond (1818-1896) Anton Dohrn (1840-1909): Briefwechsel.* Berlin: Springer.

Groessens, Eric, and Marie-Claire Groessens-Van Dyck. 2001. "De geologie." In *Geschiedenis van de wetenschappen in België, 1815-2000,* edited by Robert Halleux, Geert Vanpaemel, Jan Vandersmissen, and Andrée Despy-Meyer. Vol. 2, 269-288. Brussels: Dexia/La Renaissance du Livre.

Gudermann, Rita. 2001. "Der Take-off der Landwirtschaft im 19. Jahrhundert und seine Konzequenzen für Umwelt und Gesellschaft." In *Agrarmodernisierung und ökologische Folgen: Westfalen vom 18. Bis zum 20. Jahrhundert,* edited by Karl Ditt, Rita Gudermann, and Norwich Rüße, 47-83. Paderborn: Ferdinand Schöningh.

Günther, Franz. 1984. "Pfarrer als Wissenschaftler." In *Das evangelische Pfarrhaus: eine Kultur- und Sozialgeschichte,* edited by Martin Greiffenhagen, 277-294. Stuttgart: Kreuz.

Guttstadt, Albert. 1886. *Die naturwissenschaftlichen und medicinischen Staatsanstalten Berlins: Festschrift für die 59. Versamml. dt. Naturforscher und Aerzte.* Berlin: Hirschwald.

Haberling, Wilhelm. 1924. *Johannes Müller: Das Leben des Rheinischen Naturforschers.* Leipzig: Akademische Verlaggeselschaft.

Hachet-Souplet, Pierre. 1909. "Un projet de transformation de la ménagerie du Jardin des plantes." *L'illustration* 66: 178-179.

Haeckel, Ernst. 1866. *Generelle Morphologie der Organismen.* 2 vols. Berlin: Reimer.

Haffer, Jürgen. 2003. "Gruppenbilder von frühen Jahresversammlungen der DOG (1872-1900)." *Journal of Ornithology* 144: 116-123.

———. 2008. *Ornithology, Evolution, and Philosophy: The Life and Science of Ernst Mayr 1904-2005.* Berlin: Springer.

Haffer, Jürgen, Erich Rutschke, and Klaus Wunderlich. 2000. *Erwin Stresemann (1889-1972): Leben und Werk eines Pioniers der wissenschaftlichen Ornithologie.* Halle: Deutsche Akademie der Naturforscher Leopoldina.

Haggerty, Melvin Everett. 1912. "La nouvelle pyschologie animale, by Georges Bohn." *Journal of Philosophy, Psychology and Scientific Methods* 9: 164-166.

Hamlin, Christopher. 1986. "Robert Warington and the Moral Economy of the Aquarium." *Journal of the History of Biology* 19: 131-153.

Hamoir, Gabriel. 2002. *La révolution évolutioniste en Belgique: Du fixiste Pierre-Joseph Van Beneden ... à son fils darwiniste Edouard.* Liège: Les Éditions de l'Université de Liège.

Harwood, Jonathan. 1987. "National Styles in Science: Genetics in Germany and the United States between the World Wars." *Isis* 78: 390-414.

———. 1996. "Weimar Culture and Biological Theory: A Study of Richard Woltereck (1877-1944)." *History of Science* 43: 347-377.

Heincke, Friedrich. 1893. "Die Biologische Anstalt auf Helgoland und ihre Tätigkeit im Jahre 1893." *Wissenschaftlichen Meeresuntersuchungen* 1: 1-33.

Helmholtz, Hermann, Rudolf Virchow, and Edmond du Bois-Reymond. 1980 [1879]. "Die zoologische Station in Neapel. Eingabe an den Reichstag Privatdruck." In *Anton Dohrn und die Zoologische Station Neapel,* edited by Hans-Reiner Simon, 49-51. Frankfurt am Main: Edition Erbrich.

Hensen, Victor. 1887. "Über die Bestimmung des Planktons oder des im Meere treibenden Materials ans Pflanzen und Thieren." 5. *Bericht der Kommission zur wissenschaftlichen Untersuchung der Deutsche Meere bei Kiel* 1-8.

Herdman, William Abbott. 1901. "The Greatest Biological Station in the World." *Popular Science Monthly* 59: 419-429.

————. 1904. *A History of Science*. New York: Harper and Brothers. Vol. 5.

Herman, Otto. 1911. "Kurze Übersicht der Organization und Arbeit der Königlich Ungarischen Ornithologischen Zentrale." In *Verhandlungen des V. Internationalen Ornithologen-Kongresses. Berlin 1910*, edited by Herman Schalow, 133–143. Berlin: Deutsche Ornithologische Gesellschaft.

Hertwig, Oskar. 1906. *Allgemeine Biologie*. 2nd ed. Jena: Gustav Fischer.

Herzig, Rebecca M. 2005. *Suffering for Science: Reason and Sacrifice in Modern America*. New Brunswick, NJ: Rutgers University Press.

Heuss, Theodor. 1991. *Anton Dohrn: A Life for Science*. Berlin: Springer.

Hostyn, Norbert. 1976. "Een natte attractie uit 1894: Lebon's aquarium." *De Plate* 5: 13–14.

Houzé, Emile. 1906. *L'Aryen et l'anthroposociologie: Étude critique*. Brussels: Misch et Thron.

Howard, Leland Osian. 1912. "The Recent Progress and Present Conditions of Economic Entomology." *Proceedings of the Seventh International Zoological Congress*. Cambridge, MA: Harvard University Press, 572–600.

Hünemörder, Christian. 1995. "Ornithology on the Island of Helgoland and the Role of the Biologische Anstalt up to the Foundation of the Separate 'Vogelwarte.'" *Helgoländer Meeresuntersuchungen* 49: 125–134.

Jack, Homer A. 1940. *The Biological Field Stations of the World—A Comparative and Descriptive Study*. Master's thesis, Cornell University.

————. 1945. "Biological Field Stations of the World." *Chronica Botanica* 9: 2–73.

Janet, Pierre. 1900–1901. "Société internationale de l'institut psychique." *Bulletin de l'institut psychologique international* 1: 3–12.

Jansen, Sarah. 2002. "Den Heringen einen Pass ausstellen: Formalisierung und Genauigkeit in den Anfängen der Populationsökologie um 1900." *Berichte zur Wissenschaftsgeschichte* 25: 153–169.

————. 2003. *"Schädlinge": Geschichte eines wissenschaftlichen und politischen Konstrukts 1840–1920*. Frankfurt/Main: Campus Verlag.

Jaynes, Julian. 1969. "The Historical Origins of 'Ethology' and 'Comparative Psychology.'" *Animal Behavior* 17: 601–606.

Junge, Friedrich. 1885. *Der Dorfteich als Lebensgemeinschaft nebst einer Abhandlung über Ziel und Verfahren des naturgeschichtlichen Unterrichts*. Kiel: Lipsius & Tischer.

Juday, Chauncey. 1910. "Some European Biological Stations." *Transactions of the Wisconsin Academy of Sciences, Arts and Letters* 16: 1257–1277.

Kemna, Adolphe. 1897. *P. J. Van Beneden: La vie et l'oeuvre d'un zoologiste*. Antwerp: Imprimerie J.-E. Buschman.

Kingsland, Sharon E. 2005. *The Evolution of American Ecology, 1890–2000*. Baltimore: Johns Hopkins University Press.

————. 2010. "The Role of Place in the History of Ecology." In *The Ecology of Place: Contributions of Place-Based Research to Ecological Understanding*, edited by Ian Billick and Mary V. Price, 11–39. Chicago: University of Chicago Press.

Kinsey, Darin. 2006. "'Seeding the Water as the Earth': The Epicenter and Peripheries of a European *Aqua*cultural Revolution." *Environmental History* 11: 527–566.

Kinzelbach, Ragnar. 2005 *Das Buch vom Pfeilstorch*. Marburg an der Lan: Bassiliskenpresse.

Klös, Heinz-Georg, Hans Frädich, and Ursula Klös. 1994. *Die Arche Noah an der Spree: 150 Jahre Zoologischer Garten Berlin. Eine tiergärtnische Kulturgeschichte von 1844–1994*. Berlin: FAB-Verlag.

Knight, Legh. 1868. *Tonic Bitters: A Novel*. London: Chapman and Hall.

Kofoid, Charles A. 1910. *The Biological Stations of Europe*. Washington, DC: Government Printing Office.

Kohler, Robert. 2002. *Landscapes and Labscapes: Exploring the Lab-Field Border in Biology*. Chicago: University of Chicago Press.

———. 2002. "Place and Practice in Field Biology." *History of Science* 40: 189–210.

———. 2006. *All Creatures: Naturalists, Collectors and Biodiversity, 1850–1950*. Princeton: Princeton University Press.

———. 2008. "Lab History—Reflections." *Isis* 99: 761–768.

Kohlstedt, Sally G. 1987. "International Exchange and National Style: A View of Natural History Museums in the United States, 1850–1900." In *Scientific Colonialism: A Cross-Cultural Comparison*, edited by Nathan Reingold and Marc Rothenberg, 167–190. Washington, DC: Smithsonian Institution Press.

Koller, Gottfried. 1958. *Johannes Müller 1801–1858*. Stuttgart: Wissenschaftliche Verlagsgesellschaft M.B.H.

Kossert, Andreas. 2007. *Ostpreussen: Geschichte und Mythos*. Pössneck: Pantheon.

Köstering, Susanne. 2003. *Natur zum Anschauen: Das Naturkundemuseum des deutschen Kaiserreichs 1871–1914*. Köln: Böhlau Verlag.

Kräcker, Julius. 1880. *Etwas mehr Licht über die Ursachen des Nothstandes in Oberschlesien*. Breslau: Verl. der Schlesischen Volksbuchh.

Krassilstchik, Isaak. 1888. "La production industrielle des parasites végétaux pour la destruction des insectes nuisibles." *Bulletin biologique de la France et de la Belgique* ser. 3, 1: 460–472.

Krauße, Erika. 1995. "Haeckel: Promorphologie und 'evolutionistische' ästhetische Theorie. Konzept und Wirkung." In *Die Rezeption von Evolutionstheorien im 19. Jahrhundert*, edited by Eve-Marie Engels, 347–394. Frankfurt am Main: Suhrkamp.

Kremer, Richard L. 1992. "Building Institutes for Physiology in Prussia, 1836–1846: Contexts, Interests and Rhetoric." In *The Laboratory Revolution in Medicine*, edited by Andrew Cunningham and Perry Williams, 72–109. Cambridge: Cambridge University Press.

Kretschmann, Carsten. 2006. *Räume öffnen sich: Naturhistorische Museen im Deutschland des 19. Jahrhundert*. Berlin: Akademie Verlag.

Kruuk, Hans. 2003. *Niko's Nature: The Life of Niko Tinbergen and His Science of Animal Behavior*. Oxford: Oxford University Press.

Lameere, August. 1896–1897. "Laboratoire de biologie ambulant de l'université de Bruxelles." *Revue de l'Université de Bruxelles* 2: 788–789.

———. 1927. "L'Institut zoologique Torley-Rousseau." *Receuil de l'Institut zoologique Torley-Rousseau* 1: 5–9.

———. 1938. "Guillaume Severin (1862–1938)." *Bulletin et annales de la societé entomologique de Belgique* 78: 313–314.

Lampert, Kurt. 1899. *Das Leben der Binnengewässer*. Leipzig: Tauchnitz.

Lange, Ulrich. 1996. "Modernisierung der Infrastruktur." In *Geschichte Schleswichs Holsteins: Von den Anfängen bis zur Gegenwart*, edited by Ulrich Lange, 346–367. Wachholtz Verlag: Neumünster.

Lankester, Edwin Ray. 1888. "Zoology." In *Encyclopaedia Britannica*. Vol. 24, 799–820. New York: Henry G. Allen.

Largent, Mark A. 1999. "Bionomics: Vernon Lyman Kellogg and the Defence of Darwinism." *Journal of the History of Biology* 32: 465–488.

Larsen, Anne. 1996. "Equipment for the Field." In *Cultures of Natural History*, edited by Nick Jardine, James A. Secord, and Emma C. Spary, 358–377. Cambridge: Cambridge University Press.

Laubichler, Manfred D. 2006. "Allgemeine Biologie als selbständige Grundwissenschaft und die allgemeine Grundlagen des Lebens." In *Der Hochsitz des Wissens. Das Allgemeine als wissenschaftlicher Wert*, edited by Manfred Hagner and Manfred D. Laubichler, 185–205. Zürich: Diaphanes Verlag.

Lauterborn, Robert. 1902. "Das Projekt einer schwimmenden biologischen Station zur Erforschung des Tier- und Pflanzenlebens unserer Ströme." In *Verhandlungen des V. internationalen Zoologen-Kongresses Berlin*, 307-312. Jena: Fischer.

La Vergata, Antonello. 1996. "Lamarckisme et solidarité." *Asclepio: Revista di historia de la medicina y de la sciencia* 1: 273-288.

Le Dantec, Félix. 1909. "Jules Bonnier, directeur-adjoint du laboratoire d'évolution des êtres organisés et de la station zoologique de Wimereux, 1859-1908." *Bulletin scientifique de la France et de la Belgique* 42: lxxv-lxxviii.

Lenoir, Timothy. 1982. *The Strategy of Life: Teleology and Mechanics in Nineteenth-Century German Biology*. Boston: Reidel.

———. 1992. "Laboratories, Medicine and Public Life in Germany 1830-1849: Ideological Roots of the Institutional Revolution." In *The Laboratory Revolution in Medicine*, edited by Andrew Cunningham and Perry Williams, 14-71. Cambridge: Cambridge University Press.

Lenz, Friedrich. 1936. "Limnologische Laboratorien." In *Handbuch der biologischen Arbeitsmethoden*, edited by Emil Abderhalden. Vol. 2, 1287-1368. Berlin: Urban und Schwarzenberg.

Leps, Günther. 1998. "Ökologie und Ökosystemforschung." *Geschichte der Biologie: Theorien, Methoden, Institutionen, Kurzbiographien*, edited by Ilse Jahn, 601-619. Jena: Gustav Fischer.

———. 2006. "Karl August Möbius, ein Klassiker des ökologisches Denkens" In *Zum Biozönose-Begriff: Die Auster und die Austernwirtschaft*, xxxv-lxiv. Frankfurt am Main: Verlag Harri Deutsch.

Lesch, John E. 1984. *Science and Medicine in France: The Emergence of Experimental Physiology 1790-1855*. Cambridge: Harvard University Press.

Lestage, Johannes Antoine. 1921. "Le Dr. Ernest Rousseau: Sa vie-son œuvre." *Les naturalistes belges* 2: 261-279.

Limoges, Camille. 1980. "The Development of the Muséum d'Histoire Naturelle of Paris, c. 1800-1900." In *The Organization of Science and Technology in France 1808-1914*, edited by Robert Fox and George Weisz, 230-240. Cambridge: Cambridge University Press.

Livingstone, David N. 2003. *Putting Science in Its Place: Geographies of Scientific Knowledge*. Chicago: University of Chicago Press.

Lloyd, William Alford. 1876. "Aquaria: Their Present, Past and Future." *Popular Science Review* 15: 253-265.

Lo Bianco, Salvatore. 1899. *The Methods employed at the Naples Zoological Station for the Preservation of Marine Animals*. Washington: Government Printing Office.

Loeb, Jacques. 1908. *La dynamique des phénomènes de la vie*. Paris: Alcan.

Loisel, Gustave. 1905. *Exposé des titres et des travaux scientifiques (1892-1905)*. Paris: Impr. De Lahure.

———. 1907. *Projets et études sur la réorganization et l'utilisation de la ménagerie du Jardin des plantes*. Paris: Ed. de La Revue des idées.

———. 1908. "Rapport sur une mission scientifique dans les jardins et établissements zoologiques publics et privés des États-Unis et du Canada, avec conclusions générales sur les jardins zoologiques." *Nouvelles archives des missions scientifiques et littéraires* 26: 217-407.

Loison, Laurent. 2008. *Les notions de plasticité et d'hérédité chez les néo-lamarckiens français (1879-1946): Éléments pour une histoire du transformisme français*. PhD diss., Université de Nantes.

———. 2011. "French Roots of French Neo-Lamarckisms, 1879-1985." *Journal of the History of Biology* 44: 713-744.

Löns, Hermann. 1909. *Mümmelman: Ein Tierbuch*. Hannover: Spontholz.

———. 1910. "Magenuntersuchung und Beringung." *Deutsche Jäger-Zeitung* 55: 440-441.

Lord, Jeffrey. 2005. "From Metchnikoff to Monsanto and Beyond: The Path of Microbial Control." *Journal of Invertebrate Pathology* 89: 19-29.

Lorenzen-Schmidt, Klaus-Joachim. 1996. "Bevölkerungsentwicklung." In *Geschichte Schleswichs Holsteins: Von den Anfängen bis zur Gegenwart*, edited by Ulrich Lange, 341–345. Wachholtz Verlag: Neumünster.

———. 1996. "Neuorientierung auf den deutschen Wirtschaftsraum—Wirtschaftliche Entwicklung 1864–1918." In *Geschichte Schleswichs Holsteins: Von den Anfängen bis zur Gegenwart*, edited by Ulrich Lange, 385–399. Wachholtz Verlag: Neumünster.

Lowe, Philip D. 1976. "Amateurs and Professionals: The Institutional Emergence of British Plant Ecology." *Journal of the Society for the Bibliography of Natural History* 7: 517–535.

MacKenzie, John, 1988. *The Empire of Nature*. Manchester: Manchester University Press.

MacLeod, Julius. 1882. "La station zoologique de Naples." *Annales de la Société de médecine de Gand* 52: 13–21.

Maienschein, Jane. 1985. "First Impressions: American Biologists at Naples." *Biological Bulletin* 168: 187–191.

———. 1986. "Arguments for Experimentation in Biology." *Proceedings of the Biennial Meeting of the Philosophy of Science Association* 2: 180–195.

———. 1989. *100 Years Exploring Life, 1888–1988: The Marine Biological Laboratory at Woods Hole*. Boston: Jones & Bartlett.

———. 1991. *Transforming Traditions in American Biology, 1880–1915*. Baltimore: Johns Hopkins University Press.

———. 1994. "'It's a Long Way from Amphioxus': Anton Dohrn and Late Nineteenth Century Debates about Vertebrate Origins." *History & Philosophy of the Life Sciences* 16: 465–478.

Marey, Etienne-Jules. 1890. "Locomotion dans l'eau." *La nature: Revue des sciences et de leurs applications aux arts et à l'industrie* 18: 375–378.

Marinescu, Alexandru, and Gheorghe Bratescu. 1992. "Une controverse de 1872 sur la définition de la science expérimentale: La polémique entre Claude Bernard et Henri de Lacaze-Duthiers." *Travaux du Muséum d'Histoire Naturelle "Grigore Antipa"* 32: 507–513.

Marmin, Nicolas. 2001. "Métapsychique et psychologie en France (1880–1940)." *Revue d'histoire des sciences humaines* 1: 145–175.

Massart, Jean. 1893. "La biologie et la végétation sur le littoral belge." *Bulletin de la Société royale de botanique de Belgique* 32: 7–43.

———. 1912. "Les naturalistes actuels et l'étude de la nature." *Bulletins de l'Académie royale belge des sciences, des lettres et des beaux-arts* ser. 4, 11: 944–965.

———. 1912. *Pour la protection de la nature*. Brussels: Lamertin.

———. 1921. *Éléments de biologie générale et de botanique*. 2 vols. Brussels: Lamertin.

———. 1922. "La biologie des inondations de l'Yser et la flore des ruines de Nieuport." *Recueil de l'Institut botanique Léo Errera* 10: 411–430.

Massart, Jean, and Émile Vandervelde. 1893. "Parasitisme organique et parasitisme social." *Bulletin scientifique de la France et de la Belgique* 25: 227–294.

Matagne, Patrick. 1999. *Aux origines de l'écologie: Les naturalistes en France de 1800 à 1914*. Paris: Éditions du CTHS.

———. 2011. "The French Tradition in Ecology: 1820–1950." In *Ecology Revisited: Reflecting on Concepts, Advancing Science*, edited by Astrid Schwarz and Kurt Jax, 287–306. Dordrecht: Springer.

McElligot, Anthony. 2001. *The German Urban Experience: Modernity and Crisis*. New York: Routledge.

McIntosh, Robert P. 1985. *The Background of Ecology: Concept and Theory*. Cambridge: Cambridge University Press.

Mckenzie, Callum. 2005. "'Sadly Neglected'-Hunting and Gendered Identities: A Study in Gender Construction." *International Journal of the History of Sport* 22: 545–562.

McNeely, Ian F. 2002. *"Medicine on a Grand Scale": Rudolf Virchow, Liberalism, and the Public Health*. London: Wellcome Trust.

Ménégaux, Auguste. 1905. "Le laboratoire maritime de Wimereux." *Bulletin de l'Institut général psychologique* 6: 459-485.

Menuge-Wacrenier, Raymonde. 1986. *La Côte d'Opale à la Belle Époque.* Dunekerque: Westhoek.

Meyer, Heinrich, and Karl Möbius. 1865. *Die Fauna der Kieler Bucht.* Vol. 1, *Die Hinterkiemer oder Opisthobranchia der Kieler Bucht.* Leipzig: Wilhelm Engelmann.

———. 1872. *Die Fauna der Kieler Bucht.* Vol. 2, *Die Prosobranchia und Lamellibranchia, nebst einem supplement zu den Opisthobranchia.* Leipzig: Wilhelm Engelmann.

Michelant, Louis. 1880. *Boulogne-sur-Mer, Berck, Calais, Dunkerque.* Paris: Hachette.

Miles, Ashley. 1982-1983. "Reports by Louis Pasteur and Claude Bernard on the Organization of Scientific Teaching and Research." *Notes and Records of the Royal Society of London* 37: 101-118.

Mills, Eric L. 1989. *Biological Oceanography: An Early History, 1870-1960.* Ithaca: Cornell University Press.

Mills, Eric L. 1990. "The Ocean Regarded as a Pasture: Kiel, Plymouth and the Explanation of the Marine Plankton Cycle, 1887 to 1935." *Deutsche Hydrographische Zeitschrift* 22: 20-29.

———. 2009. *The Fluid Envelope of Our Planet: How the Study of Ocean Currents Became a Science.* Toronto: University of Toronto Press.

Minot, Charles Sedwick. 1884. "The Laboratory in Modern Science." *Science* 3: 172-174.

Mitman, Gregg. 1999. *Reel Nature: America's Romance with Wildlife on Film.* Cambridge, MA: Harvard University Press.

Mitman, Gregg, Michelle Murphy, and Christopher Sellers, eds. 2004. *Landscapes of Exposure: Knowledge and Illness in Modern Environments.* Chicago: University of Chicago Press. *Osiris* 19.

Möbius, Karl. 1877. *Die Auster und die Austernwirtschaft.* Berlin: Wiegand, Hempel & Parey.

———. 1883. "The Oyster and Oyster-Culture." In *Report for 1880,* by US Fish Commission. Washington: Government Printing Office.

Möller, Rudolf. 2004. "Notizen zur Biographie des Ornithologen Johannes Thienemann und zur Geschichte der Vogelwarte Rossitten (heute Rybačij)." *Rudolstädter Naturhistorische Schriften* 12: 147-194.

Monroy, Alberto, and Christiane Groeben. 1985. "The 'New' Embryology at the Zoological Station and at the Marine Biological Laboratory." *Biological Bulletin* 168: 35-43.

Müller, Irmgard. 1973. "Der 'Hydriot' Nikolai Kleinenberg oder Spekulation und Beobachtung." *Medizinhistorisches Journal* 8: 131-153.

———. 1975. "Die Wandlung embryologischer Forschung von der deskriptiven zur experimentellen Phase unter dem Einfluss der Zoologischen Station in Neapel." *Medizinhistorisches Journal* 10: 191-204.

———. 1996. "The Impact of the Zoological Station in Naples on Developmental Physiology." *International Journal of Developmental Biology* 10: 103-111.

Nash, Roderick. 1967. *Wilderness and the American Mind.* New Haven, CT: Yale University Press.

Naylor, Simon. 2005. "Introduction: Historical Geographies of Science—Places, Contexts, Cartographies." *British Journal for the History of Science* 38: 1-12.

Nöthlich, Rosemarie, Nadine Wetzel, Uwe Hoßfeld, and Lennart Olsson. 2006. "'Ich acquirirte das Schwein sofort, ließ nach dem Niederstechen die Pfoten abhacken u. schickte dieselben an Darwin'—Der Briefwechsel von Otto Zacharias mit Ernst Haeckel (1874-1898)." *Annals of the History and Philosophy of Biology* 11: 177-248.

Nunn Whitman, Emily. 1886. "The Zoological Station at Naples." *Century Illustrated Monthly Magazine* September: 798-884.

Nyhart, Lynn K. 1995. *Biology Takes Form: Animal Morphology and the German Universities, 1800-1900.* Chicago: University of Chicago Press.

———. 1996. "Natural History and the 'New' Biology." In *Cultures of Natural History,* edited by Nick Jardine, James A. Secord, and Emma C. Spary, 426-443. Cambridge: Cambridge University Press.

————. 2004. "Science, Art and Authenticity in Natural History Museums." In *Models: The Third Dimension of Science*, edited by Soroya de Chadarevian and Nick Hopwood, 207-335. Stanford: Stanford University Press.

————. 2009. *Modern Nature: The Rise of the Biological Perspective in Germany*. Chicago: University of Chicago Press.

————. 2012. "Voyaging and the Scientific Expedition Report, 1800-1940." In *Science in Print: Essays on the History of Science and the Culture of Print*, edited by Greg Downey, Rima Apple, and Christine Pawley, 65-86. Madison: University of Wisconsin Press.

Nys, Liesbet. 2008. "Aspirations to Life: Pleas for New Forms of Display in Belgian Museums around 1900." *Journal of the History of Collections* 20: 113-126.

Olalquiaga, Celeste. 1999. *The Artificial Kingdom: A Treasury of the Kitsch Experience*. London: Bloomsbury.

Ophir, Adi, and Steven Shapin. 1991. "The Place of Knowledge: A Methodological Survey." *Science in Context* 4: 3-21.

Opitz, Donald L. 2006. "'This House Is a Temple of Research': Country-House Centres for Science." In *Repositioning Victorian Sciences: Shifting Centres in Nineteenth-Century Scientific Thinking*, edited by David Clifford, Elisabeth Wadge, Alex Warwick, and Martin Willis, 143-153. London: Anthem.

Otis, Laura. 2007. *Müller's Lab*. Oxford: Oxford University Press.

Outram, Dorrinda. 1984. *Georges Cuvier: Vocation, Science and Authority in Post-Revolutionary France*. Manchester: Manchester University Press.

————. 1995. "New Spaces in Natural History." In *Cultures of Natural History*, edited by Nick Jardine, Jim Secord, and Emma Sparry, 249-265. Cambridge: Cambridge University Press.

Papadopoulos, Alex G. 1996. *Urban Regimes and Strategies: Building Europe's Central Executive District in Brussels*. Chicago: University of Chicago Press.

Paul, Harry. 1985. *From Knowledge to Power: The Rise of the Science Empire in France, 1860-1939*. Cambridge: Cambridge University Press.

Pauly, Philip J. 1988. "Summer Resort and Scientific Discipline: Woods Hole and the Structure of American Biology, 1882-1925." In *The American Development of Biology*, edited by Ronald Rainger, Keith R. Benson, and Jane Maienschein, 121-150. Philadelphia: University of Pennsylvania Press.

————. 2000. *Biologists and the Promise of American Life: From Meriwether Lewis to Alfred Kinsey*. Princeton: Princeton University Press, 2000.

Pelseneer, Paul. 1902-1903. "La morale de la science et la morale de l'Eglise." *Bulletin des travaux du Suprême conseil de Belgique* 46: 34-47.

————. 1908. "Alfred Giard (1846-1908) in memoriam." *Annales de la Société royale zoologique et malacologique de Belgique* 43: 220-228.

————. 1935. *Essai d'éthologie zoologique d'après l'étude des mollusques*. Brussels: Académie royale de Belgique.

Perrier, Edmond. 1888. "Le laboratoire maritime du Muséum d'histoire naturelle." *La nature: Revue des sciences et de leurs applications aux arts et à l'industrie* 16: 186-188.

Persell, Stuart M. 1999. *Neo-Lamarckism and the Evolution Controversy in France*. Lewiston: Edwin Mellen Press .

Phillips, Denise. 2003. "Friends of Nature: Urban Sociability and Regional Natural History in Dresden, 1800-1850." In *Science in the City*, edited by Sven Dierig, Jens Lachmund, and Andrew Mendelsohn. *Osiris* 18: 43-59.

Pickstone, John V. 2000. *Ways of Knowing: A New History of Science, Technology and Medicine*. Manchester: Manchester University Press.

Pitte, Jean-Robert. 2003. *Histoire du paysage français: Du préhistoire à nos jours*. Paris: Tallandrier.

Potthast, Thomas. 2006. "Historische und ökologietheoretische Perspektiven auf Karl August Möbius Schrift 'Die Auster und die Austernwirtschaft.'" In *Zum Biozönose-Begriff: Die Auster und die Austernwirtschaft*, vii-xxxiv. Frankfurt am Main: Verlag Harri Deutsch.

Prouho, Henri. 1886. "Le laboratoire Arago." *La nature: Revue des sciences et de leurs applications aux arts et à l'industrie* 14: 97-99.

Quetelet, Adolphe. 1846. *Letters addressed to H.R.H. the Grand Duke of Saxe Coburg and Gotha on the Theory of Probabilities as applied to the Moral and Political Sciences*. London: C. & R. Layton.

Quintart, Alain. 1990. "L'institut royal des sciences naturelles de Belgique." In *La dynastie et la culture en Belgique*, edited by Herman Balthazar and Jean Stengers, 325-330. Antwerp: Mercator Fonds.

Rabaud, Etienne. 1904. "Observations sur les manifestations mentales chez les oiseaux." *Bulletin de l'institut général psychologique* 4: 438-443.

———. 1912. "Éthologie et comportement de diverses larves endophytes." *Bulletin scientifique de la France et de la Belgique* 46: 1-28.

———. 1919. "Le domaine et la méthode de la biologie générale." *Revue philosophique de la France et de l'étranger* 44: 1-18.

———. 1925. *Les phénomènes de convergence en biologie*. Paris: PUF.

———. 1935. *Titres et travaux scientifiques*. Paris: PUF.

Raby, Megan. 2013. *Making Biology Tropical: American Science in the Caribbean, 1898-1963*. PhD diss., University of Wisconsin-Madison.

Rahir, Edmond. 1906. "Le nouvel aquarium de Bruxelles." *La nature: Revue des sciences et de leurs applications aux arts et à l'industrie* 34: 355-357.

Raveret-Watel, Casimir. 1889. "La station aquicole de Boulogne-sur-Mer." *Revue des sciences naturelles appliqués* ser. 4, 6: 925-929.

Ray, John. 1979. "The Application of Science to Industry." In *The Organization of Knowledge in Modern America, 1860-1920*, edited by Alexandra Oleson and John Voss, 249-268. Baltimore: Johns Hopkins University Press.

Rehbock, Philip. 1979. "The Early Dredgers: 'Naturalizing' in British Seas, 1830-1850." *Journal of the History of Biology* 12: 293-368.

———. 1980. "The Victorian Aquarium in Ecological and Social Perspective." In *Oceanography: The Past*, edited by M. Sears and D. Merriman, 522-539. New York: Springer.

Reingold, Nathan, and Ida H. Reingold. 1981. *Science in America: A Documentary History 1900-1939*. Chicago: University of Chicago Press.

Reynouard, Jean. 1909. "Besse d'aujourd'hui, Besse d'autrefois." *Annales de la station limnologique de Besse* 1: 1-14.

Richard, Jules. n.d. *Le Musée Océanographique de Monaco*. Monaco: Le Musée Océanographique.

Richards, Robert. 2002. *The Romantic Conception of Life: Science and Philosophy in the Age of Goethe*. Chicago: University of Chicago Press.

———. 2007. *The Tragic Sense of Life: Ernst Haeckel and the Struggle over Evolutionary Thought*. Chicago: University of Chicago Press.

Rieper, Adolf. 1911. "Über die Ferienkurse an der Biologische Station zu Plön." *Zoologischer Anzeiger* 37: 30-32.

———. 1911. "Antwort auf die Entgegnung von Prof. Zacharias." *Zoologischer Anzeiger* 37: 319-320.

———. 1911. "Nochmals die Plöner Ferienkurse." *Zoologischer Anzeiger* 37: 575-576.

Ringer, Fritz. 1969. *The Decline of the German Mandarins: The German Academic Community, 1890-1933*. Cambridge, MA: Harvard University Press.

———. 2000. "*Bildung* and Its Implications in the German Tradition, 1890-1930." In *Toward a Social History of Knowledge: Collected Essays*, edited by Fritz Ringer, 194-212. New York: Berghahn.

Robeyns, Guy. 2009. "Aquarium und Museum für Fischzucht, 1906-1937, in Brüssel, Belgien." *Der Zoologische Garten* 78: 300-313.

Rörig, Georg. 1900. *Magenuntersuchungen von land- und forstwirtschaftlich wichtigen Vögeln*. Berlin: Gesundheitsamt.

Rossmässler, Emil Adolf. 1857. *Das Süsswasser-Aquarium: Eine Anleitung zur Herstellung und Pflege desselben*. Leipzig: Hermann Mendelssohn.

Rothfels, Nigel. 2002. *Savages and Beasts: The Birth of the Modern Zoo*. Baltimore: Johns Hopkins University Press.

Roule, Louis. 1923. "Les musées régionaux d'histoire naturelle et leur rôle dans l'enseignement publique." *Revue scientifique (revue rose)* 61: 129-136.

Rousseau, Ernest. 1906. "Avant-Propos." *Annales de biologie lacustre* 1: ix-xi.

―――. 1906. "La station biologique d'Overmeire." *Annales de biologie lacustre* 1: 311-320.

―――. 1914. *Rapport sur les bassins d'élevage pour Salmonides*. Brussels: Commission de Pisciculture.

―――. 1915. *Les poissons d'eau douce indigènes et acclimatisés de la Belgique*. Brussels: Station biologique d'Overmeire.

―――. 1917. "L'Aquarium et le Musée de Pisciculture de Bruxelles." In *Guide illustré de Bruxelles*, edited by Guillaume Des Marez. Vol. 2, 285-297. Brussels: Reynaert.

―――. 1920. "Le laboratoire d'Overmeire." *Les naturalistes belges* 1: 25-28.

―――. n.d. *Notes sur la pollution des eaux: La dilution limite*. No place and publisher mentioned.

Rousseau, Ernest, Johannes-Antoine Lestage, and Henri Schouteden. 1921. *Les larves et nymphes aquatiques des insectes d'Europe*. Brussels: Office de Publicité.

Rozwadowski, Helen M. 2002. *The Sea Knows No Boundaries: A Century of Marine Science under ICES*. Seattle: University of Washington Press.

―――. 2005. *Fathoming the Ocean: The Discovery and Exploration of the Deep Sea*. Cambridge, MA: Belknap.

Ruttner, Franz. 1933. "Ein mobiles Laboratorium für limnologische Untersuchungen." *Internationale Revue der gesamten Hydrobiologie und Hydrographie* 29: 148-154.

Sand, René. 1898. "Les laboratoires maritimes de zoologie." *Revue de l'Université de Bruxelles* 3: 23-47,121-151, and 203-235.

Sauvage, Henri-Émile. 1883. "La station zoologique 'volante.'" *La nature: Revue des sciences et de leurs applications aux arts et à l'industrie* 11: 225-226.

Schaffer, Simon. 1998. "Physics Laboratories and the Victorian Country House." In *Making Space for Science: Territorial Themes in the Shaping of Knowledge*, edited by Crosbie Smith and Jon Agar, 149-180. Basingstoke: Macmillan.

Schalow, Herman, ed. 1911. *Verhandlungen des V. Internationalen Ornithologen-Kongresses. Berlin 1910*. Berlin: Deutsche Ornithologische Gesellschaft.

Schmeil, Otto. 1893. "Die zoologische Station zu Rovigno." *Zoologischer Anzeiger* 45: 401-404.

Schmidt, Hermann, and Georg Blohm. 1978. *Die Landwirtschaft von Ostpreussen und Pommeren. Geschichte Leistung und Eigenart der Landwirtschaft in den ehemals ostdeutschen Landesteilen seit dem Kriege 1914/18 und bis Ende der dreissiger Jahre*. Marburg: J.-G. Herder Institut.

Schmidtlein, Richard. 1879. "Beobachtungen über die Lebensweise einiger Seethiere innerhalb der Aquarien der zoologischen Station." *Mittheilungen aus der Zoologischen Station zu Neapel* 1: 1-27.

―――. 1880. *Leitfaden für das Aquarium der zoologischen Station zu Neapel*. Leipzig: W. Engelmann.

Schmitt, François. 2012. "Les deux laboratoires de zoologie maritime du Portel de 1888 à 1942." *Le Portel Notes et Documents* 31: 91-103.

Schmoll, Friedemann. 2005. "Indication and Identification: On the History of Bird Protection in Germany, 1800-1918." In *Germany's Nature. Cultural Landscapes and Environmental History*, edited by Thomas Lekan and Thomas Zeller, 161-182. New Brunswick, NJ: Rutgers University Press.

Schneider, Daniel. 2000. "Local Knowledge, Environmental Politics, and the Founding of Ecology in the United States: Stephen Forbes and 'The Lake as *Microcosm*' (1887)." *Isis* 91: 681–705.

Schneller, Gerhard. 1993. *Das Werk August Thienemanns. Die Theoretische Begründung und Entwicklung der ökologischen Limnologie und allgemeinen Ökologie zur eigenständigen Wissenschaft.* Frankfurt: Peter Lang.

Schofer, Laurence. 1975. *The Formation of a Modern Labor Force, Upper Silesia, 1865–1914.* Berkeley: University of California Press.

Schouteden-Wéry, Joséphine. 1913. *Excursions scientifiques. II. En Brabant.* Brussels: Lamertin.

Schüz, Ernst. 1930. "XXV. Bericht der Vogelwarte Rossitten." *Der Vogelzug* 1: 105–113.

———. 1932. "Vogelwarte Rossitten." *Königsberger Allgemeinen Zeitung*, 10 July.

———. 1934. "Das Vogelwarte-Museum im Dienst am Heimat und Volk." *Naturschutz: Monatschrift für alle Freunde der deutsche Heimat* 15: 214–218.

———. 1938. "Johannes Thienemann zum Gedächtnis." *Journal für Ornithologie* 86: 466–483.

———. 1952. *Vom Vogelzug: Grundriss der Vogelzugskunde.* Frankfurt/Main: P. Schöps.

Schwarz, Astrid. 2000. *Frühe Ökologie im wissenschaftlichen und kulturellen Kontext: Oszillation dreier Basiskonzepte unter besonderer Berücksichtigung der aquatischen Ökologie.* PhD diss., Technischen Universität München.

———. 2003. *Wasserwüste—Mikrokosmos—Ökosystem: Eine Geschichte der Eroberung des Wasserraumes.* Freiburg: Rombach-Verlag.

———. 2009. "Rising above the Horizon: Visual and Conceptual Modulation of Space and Place." *Augenblick* 45: 36–61.

Schwarz, Astrid, and Kurt Jax. 2011. "Early Ecology in the German-Speaking World through WWII." In *Ecology Revisited: Reflecting on Concepts, Advancing Science,* edited by Astrid Schwarz and Kurt Jax, 231–275. Dordrecht: Springer.

Secord, Anne. 2011. "Coming to Attention: A Commonwealth of Observers during the Napoleonic Wars." In *Histories of Scientific Observation,* edited by Lorraine Daston and Elizabeth Lunbeck, 421–444. Chicago: University of Chicago Press.

Seligo, Arthur. 1890. *Hydrobiologische Untersuchungen.* Danzig: L. Saunier.

———. 1908. *Tiere und Pflanzen des Seenplanktons.* Stuttgart: Franckh'sche Verlagshandlung.

Severin, Guillaume. 1902. "L'invasion de l'Hylésine géante." *Bulletin de la Société centrale forestière de Belgique* 9: 145–152.

———. 1904. "Le rôle de l'entomologie en sylviculture." *Bulletin de la Société centrale forestière de Belgique* 11: 152–162.

———. 1906. "Oiseaux insectivores et insectes nuisibles." *Bulletin de la Société centrale forestière de Belgique* 13: 196–208 and 263–274.

———. 1907. "À propos d'une note sur les musées américains." *Annales de la Société royale zoologique et malacologique de Belgique* 42: 234–262.

Shapin, Steven. 1990. "'The Mind in Its Own Place': Science and Solitude in Seventeenth-Century England." *Science in Context* 4: 191–218.

Shelford, Victor E. 1913. *Animal Communities in Temperate America.* Chicago: University of Chicago Press.

Sherard, Robert Harborough. 1893. *Émile Zola: A Biographical & Critical Study.* London: Chatto & Windus.

Shinn, Terry. 1979. "The French Science Faculty System, 1808–1914: Institutional Change and Reseach Potential in Mathematics and the Physical Sciences." *Historical Studies in the Physical Sciences* 10: 271–332.

Smith, Crosbie, and Jon Agar, eds. 1998. *Making Space for Science: Territorial Themes in the Shaping of Knowledge.* New York: Macmillan.

Snowden, Frank M. 1995. *Naples in the Time of Cholera, 1884–1911.* New York: Cambridge University Press.

Söderqvist, Thomas. 1986. *The Ecologists: From Merry Naturalists to the Saviours of the Nation.* Stockholm: Almqvist og Wiksell.

Sommerfeld, Josef. 1984. *Er Flug die Besenstielkiste: Segelflieger Ferdinand Schulz.* Munich: Schildverlag.

Sorenson, Richard. 1996. "The Ship as a Scientific Instrument in the Eighteenth Century." *Osiris* 11: 221–236.

Stebbins, Robert E. 1988. "France." In *The Comparative Reception of Darwinism*, edited by Thomas Glick, 117–167. Chicago: University of Chicago Press.

Steigerwald, Joan. 2000. "The Cultural Enframing of Nature: Environmental Histories during the German Early Romantic Period." *Environment and History* 6: 451–496.

———. 2002. "Goethe's Morphology: *Urphänomene* and Aesthetic Appraisal." *Journal of the History of Biology* 35: 291–328.

Steleanu, Adrian. 1989. *Geschichte der Limnologie und ihrer Grundlagen.* Frankfurt am Main: Haag und Herchen.

Stephenson, Roger H. 2005. "'Binary Synthesis': Goethe's Aesthetic Intuition in Literature and Science." *Science in Context* 18: 553–581.

Stevens, René. 1915. *Une excursion-type dans la forêt de Soignes.* Brussels: Agence Dechenne.

Stevens, René, and Louis Vanderswaelmen. 1914. *Guide du promeneur dans la forêt de Soignes.* Brussels: G. Van Oest.

Stockmans, François. 1973. "Notice sur Victor van Straelen." *Annuaire de l'Académie royale de Belgique* 107: 1–76.

Strehlow, Harro. 1987. "Zur Geschichte des Berliner Aquariums Unter den Linden." *Zoologischer Garten* 57: 26–40.

Stresemann, Erwin. 1951. *Die Entwicklung der Ornithologie. Von Aristoteles bis zur Gegenwart.* Berlin: F. W. Peters.

Sucker, Oswald. 1880. *Der Nothstand in Oberschlesien und die Ursachen seiner Entstehung.* Breslau: s.n.

Sues, Hans-Dieter. 1997. "European Dinosaur Hunters." In *The Complete Dinosaur*, edited by James O. Farlow and M.K. Brett-Surman, 12–23. Bloomington: Indiana University Press.

Telkes, Eva, ed. 1993. *Maurice Caullery: Un biologiste au quotidien, 1868–1958.* Lyon: Presses universitaires de Lyon.

Thackray, John, and Bob Press. 2001. *The Natural History Museum: Nature's Treasurehouse.* London: Natural History Museum.

Thienemann, August. 1917. "Otto Zacharias †: Ein Nachruf." *Archiv für Hydrobiologie und Planktonkunde* 11: i–xxiv.

———. 1917. "Die wissenschaftlichen Aufgaben und die wirtschaftliche Bedeutung der biologischen Station zu Plön." *Archiv für Hydrobiologie und Planktonkunde* 11: 624–628.

———. 1918. "Lebensgemeinschaft und Lebensraum." *Naturwissenschaftliche Wochenschrift* 33: 281–303.

———. 1921. "Seetypen." *Naturwissenschaften* 9: 343–346.

———. 1927. "Zehn Jahre Hydrobiologische Anstalt Plön der Kaiser Wilhelm-Gesellschaft." *Der Naturwissenschaften* 15: 58–760.

———. 1931. "Der *Produktionsbegriff* in der Biologie." *Archiv für Hydrobiologie* 22: 16–622.

———. 1959. *Erinnerungen und Tagebuchblätter eines Biologen. Ein Leben im Dienste der Limnologie.* Stuttgart: Scheizerbart'sche Verlagsbuchhandlung.

Thienemann, Johannes. 1902. "I. Jahresbericht (1901): Der Vogelwarte Rossitten der Deutschen Ornithologischen Gesellschaft." *Journal für Ornithologie* 50: 137–209.

———. 1903. "II. Jahresbericht (1901): Der Vogelwarte Rossitten der Deutschen Ornithologischen Gesellschaft." *Journal für Ornithologie* 51: 161–231.

———. 1904. "III. Jahresbericht (1903): Der Vogelwarte Rossitten der Deutschen Ornithologischen Gesellschaft." *Journal für Ornithologie* 52: 245-295.

———. 1905. "IV. Jahresbericht (1904): Der Vogelwarte Rossitten der Deutschen Ornithologischen Gesellschaft." *Journal für Ornithologie* 53: 360-418.

———. 1906. "V. Jahresbericht (1905): Der Vogelwarte Rossitten der Deutschen Ornithologischen Gesellschaft." *Journal für Ornithologie* 54: 429-476.

———. 1907. "VI. Jahresbericht (1906): Der Vogelwarte Rossitten der Deutschen Ornithologischen Gesellschaft." *Journal für Ornithologie* 55: 481-548.

———. 1908. "VII. Jahresbericht (1907): Der Vogelwarte Rossitten der Deutschen Ornithologischen Gesellschaft." *Journal für Ornithologie* 56: 393-445.

———. 1910. "IX. Jahresbericht (1909): Der Vogelwarte Rossitten der Deutschen Ornithologischen Gesellschaft." *Journal für Ornithologie* 58: 531-676.

———. 1910. *Die Vogelwarte Rossitten der Deutschen Ornithologischen Gesellschaft und das Kenzeichnen der Vögel.* Berlin: Verlagsbuchhandlung Paul Parey.

———. 1911. "X. Jahresbericht (1910) der Vogelwarte Rossitten der Deutschen Ornithologischen Gesellschaft." *Journal für Ornithologie* 59: 620-707.

———. 1912. "X. Jahresbericht (1910) der Vogelwarte Rossitten der Deutschen Ornithologischen Gesellschaft. II. Teil." *Journal für Ornithologie* 60: 133-243.

———. 1913. "XI. Jahresbericht der Vogelwarte Rossitten 1911. II. Teil." *Journal für Ornithologie* 61: 1-64.

———. 1921. "XIX. Jahresbericht (1919): Der Vogelwarte Rossitten der Deutschen Ornithologischen Gesellschaft." *Journal für Ornithologie* 69: 1-13.

———. 1922. "XX. Jahresbericht (1920): Der Vogelwarte Rossitten der Deutschen Ornithologischen Gesellschaft." *Journal für Ornithologie* 70: 61-69.

———. 1923. "XXI. Jahresbericht der Vogelwarte Rossitten (1921)." *Journal für Ornithologie* 71: 132-158.

———. 1926. "XXIII. und XXIV. Jahresbericht der Vogelwarte Rossitten (1923 und 1924)." *Journal für Ornithologie* 74: 53-96.

———. 1927. *Etwas über das Experiment in der Vogelzugsforschung, im besonderen über ein Experiment über die Orientierung der Vögel* (Berlin: Deutsche Akademie der Wissenschaften, 1927), reprint from *Forschungen und Fortschritte: Nachrichtenblatt der deutschen Wissenschaft und Technik.*

———. 1930. *Rossitten: Drei Jahrzehnte auf der Kurischen Nehrung.* 3rd ed. Neudamm: Neumann.

———. 1931. *Vom Vogelzuge in Rossitten.* Neudamm: Neumann.

———. n.d. [1933]. *Im Lande des Vogelzuges. Für die Jugend aus Thienemanns Rossittenbuch ausgewählt von L. W. Roose.* Neudamm: Neumann.

———. 1936. "Wie die alten sungen, so zwitschern die Jungen." *Ornithologische Monatschrift* 61: 133-150 and 165-178.

———. 1939. *Das Leben unserer Vögel.* Neudamm: Verlag J. Neumann.

Thiery, Michel. [1928]. *Een bezoek aan het museum voor natuurlijke historie.* Antwerp: De Sikkel.

Thomas, Marion. 2003. *Rethinking the History of Ethology: French Animal Behavior Studies in the Third Republic (1870-1940).* PhD diss., University of Manchester.

Thomson, William. 1885. "Scientific Laboratories." *Nature* 31: 409-413.

Timmons, Tod. 2005. *Science and Technology in Nineteenth-Century America.* Westport, CT: Greenwood Press.

Tischler, Wolfgang. 1992. *Ein Zeitbild vom Werden der Ökologie.* Frankfurt: Gustav Fischer.

Tobey, Ronald C. 1981. *Saving the Prairies: The Life Cycle of the Founding School of American Plant Ecology, 1895-1955.* Berkeley: University of California Press.

Todes, Daniel. 1978. "V. O. Kovalevskii: The Genesis, Content and Reception of His Paleontological Work." *Studies in History of Biology* 2: 99-165.

Tournay, Roland. 1966. "L'histoire des naturalistes belges." *Les naturalistes belges* 47: 265–294.

Traweek, Sharon. 1988. *Beamtimes and Lifetimes: The World of High Energy Physics.* Cambridge, MA: Harvard University Press.

Trenard, Louis, and Yves-Marie Hilaire. 1999. *Histoire de Lille.* Lille: Giard.

Triboudeau, Joseph-Désiré. 1904. *Monographie agricole du Pas de Calais.* Paris: Société d'encouragement pour l'industrie nationale.

Tucker, Jennifer. 2005. *Nature Exposed: Photography as Eyewitness in Victorian Science.* Baltimore: Johns Hokins University Press.

Van Beneden, Pierre-Joseph. 1841. "Recherches sur la structure de l'oeuf dans un nouveau genre de Polype." *Bulletins de l'Académie Royale des Sciences et des Belles-Lettres de Bruxelles* 8: 89–93.

Van Bosstraeten, Truus. 2011. "Dogs and Coca-Cola: Commemorative Practices as Part of Laboratory Culture at the Heymans Institute Ghent, 1902–1970." *Centaurus: An International Journal of the History of Science and Its Cultural Aspects* 53: 1–30.

Van den Broeck, Ernest. 1882. "Une visite à la station zoologique et l'aquarium de Naples." *Annales de la Société Royale Malacologique de Belgique* 17: 3–14.

Van Dyck, Maria Jozefina. 1983. "De zoölogische verzamelingen in het collegium regium te Leuven." *Annales de la Société zoologique de Belgique* 113: 221–226.

Vanpaemel, Geert. 1996. "Gustave Dewalque en de crisis van de Belgische geologie op het einde van de negentiende eeuw." In *Actes du 13e Congrès Benelux d'Histoire des Sciences,* 171–183. Luxemburg: Centre Universitaire de Luxembourg.

Van-Praët, Michel. 2002. "La section Zoologie, témoin des restructurations de la recherche et des relations Paris-Province." In *"Par la Science, pour la Patrie": L'Association française pour l'avancement des sciences (1872–1914): Un projet politique pour une société savante,* edited by Hélène Gispert, 159–170. Rennes: Presses Universitaires de Rennes.

Van Straelen, Victor. 1936. "Discours de M. van Straelen." In *Manifestation en l'honneur de monsieur Gustave Gilson,* 22–24. Louvain: Université Catholique de Louvain.

Vetter, Jeremy. 2010. "Rocky Mountain High Science: Teaching, Research and Nature at Field Stations." In *Knowing Global Environments: New Historical Perspectives on the Field Sciences,* edited by Jeremy Vetter, 108–134. Chicago: University of Chicago Press.

———. 2011. "Introduction: Lay Participation in the History of Scientific Observation." *Science in Context* 24: 127–141.

Viguier, Camille. 1888. "La photographie microscopique à la station zoologique d'Alger." *La nature: Revue des sciences et de leurs applications aux arts et à l'industrie.* 18: 389–391.

Vivé, Anne, and Anne Versailles. 1996. *Van museum tot instituut. 150 jaar natuurwetenschappen.* Brussels: Koninklijk Belgisch Instituut voor Natuurwetenschappen.

Voigt, Max. 1904. *Die Rotatorien und Gastrotrichen der Umgebung von Plön.* Stuttgart: E. Nägele.

Vom Brocke, Bernhard. 1996. "Die Kaiser-Wilhelm-/Max-Planck-Gesellschaft und ihre Institute zwischen Universität und Akademie. Strukturprobleme und Historiographie." In *Die Kaiser-Wilhelm-/Max-Planck-Gesellschaft und ihre Institute,* edited by Bernhard vom Brocke and Hubert Laitko, 1–32. Berlin: De Gruyter.

Von Lucanus, Friedrich. 1902. "Die Höhe des Vogelzuges auf Grund aeronautischer Beobachtungen." *Journal für Ornithologie* 50: 1–9.

———. 1922. *Die Rätsel des Vogelzuges: Ihre Lösung auf experimentellem Wege durch Aeronautik, Aviatik und Vogelberingung.* Langenfalza: Beyer und Söhne.

Wagner, Patrick. 2005. *Bauern, Junker und Beamte: Lokale Herschaft und Partizipation im Ostelbien des 19. Jahrhunderts.* Göttingen: Wallstein Verlag.

Walter, Emil. 1894. "Biologie und biologische Süsswasserstationen." *Forschungsberichte aus der Biologische Station zu Plön* 2: 138–147.

———. 1903. *Die Fischerei als Nebenbetrieb des Landwirtes und Forstmannes.* Neudamm: Neumann.

Ward, Henry. 1899. "The Fresh-Water Biological Stations of the World." *Science* 9: 497–508.

Warnotte, Daniel. 1911. "Le conflit des adaptations dans l'évolution sociale." *Archives sociologiques* 2: 1–10.

Washborn, Margaret Floy. 1912. "La nouvelle pyschologie animale, by Georges Bohn." *Philosophical Review* 21: 111–112.

Wasmann, Erich. 1901. "Biologie oder Ethologie?" *Biologische Centralblatt* 21: 391–400.

Wasmund, Erich. 1929. "Luftfahrzeuge auf limnologischer Erkundung." *Arktis* 2: 41–60.

Waxweiler, Émile. 1906. *Esquisse d'une sociologie*. Brussels: Misch et Thron.

———. 1910. "Avant-propos." *Archives sociologiques* 1: 3–11.

———. 1974 [1906]. "La vie dans les phénomènes sociaux." In *Recueil de textes sociologiques d'Emile Waxweiler 1906–1914*, 17–39. Brussels: Palais des Académies.

Weindling, Paul. 1984. "Was Social Medicine Revolutionary? Rudolf Virchow and the Revolution of 1848." *Society for the Social History of Medicine* 34: 13–18.

Weisz, George. 1983. *The Emergence of Modern Universities in France, 1863–1914*. Princeton: Princeton University Press.

Welch, William H. 1896. "The Evolution of Modern Scientific Laboratories." *Nature* 54: 87–91.

Werner, Petra. 1993. *Die Gründung der Königlichen Biologischen Anstalt auf Helgoland und ihre Geschichte bis 1945*, a special issue of *Helgoländer Meeresuntersuchungen*, 47.

Wéry, Joséphine. 1908. *Excursions scientifiques: Sur le littoral belge*. Brussels: Lamertin.

Wetzel, Nadine, and Rosemarie Nöthlich. 2006. "Vom 'Homo literatus' zum 'Self-made man' der Wissenschaft: Der Werdegang des Emil Otto Zacharias (1846–1916)." *Mauritania* 19: 463–477.

Wheeler, William Morton. 1902. "'Natural History,' 'Oecology' or 'Ethology'?" *Science* 15: 971–976.

Whitman, Charles O. 1883. "The Advantages of Study at the Naples Zoological Station." *Science* 2: 92–97.

Wildfowler [Clements, Lewis]. 1876. *Shooting and Fishing Trips in England, France, Alsace, Belgium, Holland, and Bavaria*. 2 vols. London: Chapman and Hall.

———. 1877. *Shooting, Yachting, and Sea-Fishing Trips, at Home and on the Continent*. 2 vols. London: Chapman and Hall.

Wille, Robert-Jan. Forthcoming. "The Co-production of Station Morphology and Agricultural Management in the Tropics: Transformations in Botany at the Botanical Garden at Buitenzorg, Java 1880–1904." In *Life Sciences, Agriculture and the Environment: New Perspectives*, edited by Sharon Kingsland and Denise Phillips. Dordrecht: Springer.

Williams, Henry Smith. 1904–1912. *A History of Science*. 5 vols. New York: Harper & Brothers.

Williams, John Alexander. 2006. "'The Chords of the German Soul Are Tuned to Nature': The Movement to Preserve the Natural *Heimat* from the Kaiserreich to the Third Reich." *Central European History* 29: 339–384.

Wils, Kaat. 2005. *De Omweg van de wetenschap: Het positivisme en de Belgische en Nederlandse intellectuele cultuur, 1845–1914*. Amsterdam: Amsterdam University Press.

Winsor, Mary P. 1976. *Starfish, Jellyfish, and the Order of Life: Issues in Nineteenth-Century Science*. New Haven: Yale University Press.

Wocke, Max. 1880. *Regierung und Volksvertretung gegenüber dem Nothstande in Oberschlesien*. Breslau: s.n.

Wonders, Karen. 1993. *Habitat Dioramas: Illusions of Wilderness in Museums of Natural history*. Uppsala: Acta Universitatis Upsaliensis.

Wood, Harold B. 1945. "The History of Bird Banding." *Auk* 62: 256–265.

Worster, Donald. 1985. *Nature's Economy: A History of Ecological Ideas*. Cambridge: Cambridge University Press.

Yung, Emile. 1881. "Le laboratoire de zoologie maritime de Naples." *La nature: Revue des sciences et de leurs applications aux arts et à l'industrie* 9: 230–234.

Zacharias, Otto. 1880. *Die Bevölkerungs-Frage in ihrer Beziehung zu den socialen Nothständen der Gegenwart*. Hirschberg: Commisionsverlag von Aug. Heilig.

————. 1888. "Summarischer Bericht über die Aufnahme meines Vorschlags (Studium der Süsswasserfauna betr.) seitens der Fachkreise." *Biologisches Centralblatt* 8: 185-189.

————. 1888. "Vorschlag zur Gründung von zoologischen Stationen behufs Beobachtung der Süsswasser-Fauna." *Zoologischer Anzeiger* 11: 18-27.

————. 1889. "Die Aufgaben einer lacustrisch-zoologischen Station. " *Mittheilungen des West-preusischen Fischerei-Vereins* 3: 113-119.

————. 1889. "Über die lacustrisch-biologische Station am Plöner See." *Zoologischer Anzeiger* 12: 600-604.

————. 1889. "Ueber das Einsammeln von zoologischem Material in Flüssen und Seeen." In *Anleitung zur deutschen Landes- und Volksforschung,* edited by Alfred Kirchhoff, 299-328. Stuttgart: Verlag von J. Engelhorn.

————. 1889. "Ueber die Errichtung einer zoologischen Station zum Studium der Süsswasserfaune." *Entomologische Zeitschrift: Central-Organ des Entomologischen Internation. Vereins* 3: 23-25.

————. 1891. "Die Biologische Station am Plöner See." *Mittheilungen des Westpreusischen Fischerei-Vereins* 3: 105-109.

————. 1891. "Über die wissenschaftlichen Aufgaben biologischer Süsswasser-stationen." In *Die Tier- und Pflanzenwelt des Süsswassers. Einführung in das Studium derselben,* edited by Otto Zacharias. Vol. 2, 313-331. Leipzig: J. J. Weber.

————. 1891. "Vorwort." In *Die Tier- und Pflanzenwelt des Süsswassers. Einfühung in das Studium derselben,* edited by Otto Zacharias. Vol. 1, v-vi. Leipzig: J. J. Weber.

————. 1892. "Die biologische Station zu Plön." *Zoologischer Anzeiger* 15: 36-39.

————. 1892. "Die Fauna des Süsswassers in ihren Beziehungen zu der des Meeres." In *Die Tier-und Pflanzenwelt des Süsswassers. Einfühung in das Studium derselben,* edited by Otto Zacharias. Vol. 2, 295-312. Leipzig: J. J. Weber.

————. 1892. "Vorläufiger Bericht über die Thätigkeit der Biologischen Station zu Plön." *Zoologischer Anzeiger* 15: 457-460.

————. 1893. "Entgegnung auf den Artikel des Herrn Dr. C. Apstein." *Die Heimat* 3: 201-209.

————. 1894. "Beobachtungen an Plankton des Gr. Plöner Sees." *Forschungsberichte aus der Biologische Station zu Plön* 2: 91-137.

————. 1894. "Hydrobiologische Aphorismen." *Forschungsberichte aus der Biologische Station zu Plön* 2: 147-150.

————. 1895. "Vorwort." *Forschungsberichte aus der Biologische Station zu Plön* 3: v-vii.

————. 1896. "Quantitative Untersuchungen über das Limnoplankton." *Forschungsberichte aus der Biologische Station zu Plön* 4: 1-64.

————. 1896. "Vorwort." *Forschungsberichte aus der Biologische Station zu Plön* 4: v-x.

————. 1897. "Das 25-jährige Jubiläum der Zoologischen Station in Neapel." *Illustrirte Zeitung* 15: 485-488.

————. 1897. "Das Plankton der norddeutschen Binnenseen." *Die Umschau: übersicht über die Fortschritte und Bewegungen auf dem Gesamtgebiet der Wissenschaft, Technik, Literatur und Kunst* 1: 696-701.

————. 1897. "Vorwort." *Forschungsberichte aus der Biologische Station zu Plön* 5: v-vii.

————. 1898. "Ausweis über die Benützung und den Besuch der Biologische Station zu Plön in den Jahren 1892-1897." *Forschungsberichte aus der Biologische Station zu Plön* 6: 215-219.

————. 1898. "Untersuchungen über das Plankton der Teichgewässer." *Forschungsberichte aus der Biologische Station zu Plön* 6: 89-139.

————. 1898. "Vorwort." *Forschungsberichte aus der Biologische Station zu Plön* 6: v-x.

————. 1903. "F. A. Krupp als Freund und Förderer biologischer Studien." *Biologisches Centralblatt* 23: 76-84.

————. 1903. "Über die jahreszeitliche Variation von Hyalodaphnia kahlbergensis Schoedl." *Forschungsberiche aus der Biologische Station zu Plön* 10: 293-295.

———. 1904. *Die Biologische Station in Plön. Sonderabdruck aus der "Illustrirten Zeitung" Nr. 3166 vom 3. März 1904.* Leipzig: J. J. Weber.

———. 1905. "Die moderne Hydrobiologie und ihr Verhältnis zur Fischzucht und Fischerei." *Archiv für Hydrobiologie und Planktonkunde* 1: 82-108.

———. 1905. "Das Plankton als Gegenstand eines zeitgemässen biologischen Schulunterrichts." *Archiv für Hydrobiologie und Planktonkunde* 1: 246-344.

———. 1905. "Über die systematische Durchforschung der Binnengewässer und ihre Beziehung zu den Aufgaben der allgemeinen Wissenschaft vom Leben." *Forschungsberichte aus der Biologische Station zu Plön* 12: 1-34.

———. 1905. *Ueber die wissenschaftliche Bedeutung biologischer Susswasser-Stationen. Vortrag auf dem Internationalen Fischereikongress zu Wien (Juni 1905).* Plön: O. Kaven.

———. 1906. "Über die eventuelle Nützlichkeit der Begründung eines staatlichen Instituts für Hydrobiologie und Planktonkunde." *Archiv für Hydrobiologie und Planktonkunde* 2: 245-319.

———. 1907. *Das Süsswasser-Plankton: Einführung in die freischwebende Oganismenwelt unserer Teiche, Flüsse und Seebecken.* Leipzig: Teubner.

———. 1907. *Das Plankton als Gegenstand der naturkundligen Unterweisung in die Schule.* Leipzig: Theod. Thomas.

———. 1908. "Biologische Schülerübungen in die Lehrerseminarien." *Archiv für Hydrobiologie und Planktonkunde* 4: 273-275.

———. 1908. "Die staatliche Sanktion des biologischen Unterrichts." *Archiv für Hydrobiologie und Planktonkunde* 4: 233-266.

———. 1908. "Ferienkurse in Hydrobiologie und Planktonkunde an der Biologischen Station zu Plön." *Archiv für Hydrobiologie und Planktonkunde* 4: 267-272.

———. 1910. "Der Ferienkurse-Pavillion zu Plön." *Archiv für Hydrobiologie und Planktonkunde* 4: 62-82.

———. 1911. "Ein letztes Wort in Sachen meiner Plöner biologischen Ferienkurse." *Zoologischer Anzeiger* 37: 511-512.

———. 1911. "Über den speziellen Zweck und das Lehrziel der Plöner hydrobiologischen Ferienkurse." *Archiv für Hydrobiologie und Planktonkunde* 5: 271-290.

———. 1911. "Zur Entgegnung auf den Artikel des Herrn A. Rieper." *Zoologischer Anzeiger* 37: 88-94.

———. 1912. "Über chromatophile Körperchen (Parachromosomen) in den Kernen der Eimutterzellen von *Ascaris megalocephala.*" *Zoologischer Anzeiger* 40: 25-29.

———. 1912. "Zur Cytologie des Eies von *Ascaris megalocephala* (Pronuclei, gelegentliche Fusion derselben, theloide Blastomerenkerne, Chromosomen-Individualität)." *Anatomischer Anzeiger* 42: 353-384.

———. 1913. "Über den feineren Bau der Eiröhren von *Ascaris megalocephala,* insbesondere über zwei ausgedehnte Nervengeflechte in denselben." *Anatomischer Anzeiger* 43: 193-211.

———. 1917. "† Ernst Lemmermann." *Archiv für Hydrobiologie und Planktonkunde* 11: 151.

Zimmerman, Andrew. 2001. *Anthropology and Antihumanism in Imperial Germany.* Chicago: University of Chicago Press.

Zirnstein, Gottfried. 1994. *Ökologie und Umwelt in der Geschichte.* Marburg: Metropolis Verlag.

Zola, Émile. 1886. *Germinal.* Paris: La librairie illustrée.

———. 1987. *Carnets d'enquêtes: Textes établis et présentés par Henri Mitterand.* Paris: Plon.

WEBSITE

Pavé, Marc. 2005. "'To Capture or Not to Capture': French Scholars, Scientists and Alleged Fish Depletions during the Nineteenth and Twentieth Centuries." VIII Congreso de la Asociación Española de Historia Económica, Santiago de Compostela, 13-16 de septiembre. de 2005. http://www.usc.es/estaticos/congresos/histeco5/b6_pave.pdf

Index

Abel, Othenio, 112
Albert I (prince of Monaco), 43, 47-48, 202
amateurization, 203
Anzin, 101
Apstein, Carl, 136-37, 139, 143, 144
Arcachon, 27, 40, 44, 213
Audouin, Victor, 19
Audresselles, 80
aquariums: Collège de France, 20; Concarneau, 21; as hobby, 39, 126, 130, 194; as metaphor, 131, 162; Naples, 55, 62, 218; Ostend, 17; and photography, 41-42; as transportable device, 45, 53; Wimereux, 94-95, 103. *See also* public aquariums
autecology, 146, 201-2

Banyuls-sur-Mer, 42, 49, 86, 215
Bates, Walter, 56
Belgian Academy of Sciences, 180-81, 224, 226
Beneden, Edouard van, 38, 212
Beneden, Pierre-Joseph van, 17-20, 38
Berlepsch, Hans von, 155-56
Berlin: Academy of Sciences, 37, 122; Department of Statistics, 154-55; fishery station, 28; Heinroth in, 171; Müller in, 19-20, 34; natural history museum, 136, 176, 192, 237; Prussian Institute for Water, Soil and Air Hygiene, 28; public aquarium, 24-25, 40-42, 216; Thienemann in, 177; university, 19, 37, 53, 133

Bernard, Claude, 6, 34-37, 73, 89, 215
Besse-en-Chadesse, 30
Bétencourt, Alfred, 84-85, 202
Birge, Edward, 31
Blasius, Rudolf, 151
Bohn, Georges: and comparative method, 92; and comparative psychology, 106-8; and Giard, 74, 81, 86; and indoor experiment, 94-95, 202; and popular science, 102-3; in Wimereux, 84, 92, 94
Bois-Reymond, Emil du: and Dohrn, 62, 66, 215; and experiment, 34; and field stations, 34, 56, 122, 215
Bölsche, Wilhelm, 128
Bonnier, Jules, 74, 85, 91, 222
botanical gardens, 3, 30, 49, 117, 200
botany: and ecology, 6, 74-75, 89; and natural experiments, 92-93; in tropics, 7; at zoological stations, 30, 64, 118, 177
Boulogne-sur-mer, 78-81, 104
Boutan, Louis, 42
Boveri, Theodor, 51
Brandt, Karl, 136-37, 139, 144, 229
Braunsweig, 151
Breslau, 34, 53
Brussels: Institute of Sociology, 109; public aquarium, 182-83; Royal Meteorological Observatory, 149; university, 100, 102, 221, 224. *See also* Brussels natural history museum